# *Student Solutions Manual*

## to accompany

# Functions Modeling Change: A Preparation for Calculus

## Fourth Edition

**Eric Connally**

*Harvard University Extension School*

**Deborah Hughes-Hallett**

*University of Arizona*

**Andrew M. Gleason**

*Harvard University*

**et al.**

WILEY

John Wiley & Sons, Inc.

This book was set in Times Roman by the Consortium using TeX Mathematica and the package AsTeX, which as written by Alex Kasman. It was printed and bound by G&H Soho. The process was managed by Elliot Marks. Cover photo ©Patrick Zephyr/Patrick Zephyr Nature Photography

This material is based upon work supported by the National Science Foundation under Grant No. DUE-9352905. Opinions expressed are those of the authors and not necessarily those of the Foundation.

ISBN-13   978- 0-470-54735-9

Printed in the United States of America

10 9 8 7 6 5 4 3 2 1

# CONTENTS

# CHAPTER ONE

## Solutions for Section 1.1

### Skill Refresher

**S1.** Finding the common denominator we get $c + \frac{1}{2}c = \frac{2c+c}{2} = \frac{3c}{2} = \frac{3}{2}c$.

**S5.** $\left(\frac{1}{2}\right) - 5(-5) = \frac{1}{2} + 25 = \frac{51}{2}$.

**S9.** The figure is a parallelogram, so $A = (-2, 8)$.

### Exercises

**1. (a)** On the graph, the high tides occur when the graph is at its highest points. On this particular day, there were two high tides.

**(b)** The low tides occur when the graph is at its lowest points. There were two low tides on this day.

**(c)** To find the amount of time elapsed between high tides, find the distance between the two highest points on the graph. It is about 12 hours.

**5.** Appropriate axes are shown in Figure 1.1.

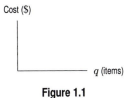

**Figure 1.1**

**9. (a)** Since $f(x)$ is 4 when $x = 0$, we have $f(0) = 4$.

**(b)** Since $x = 3$ when $f(x) = 0$, we have $f(3) = 0$.

**(c)** $f(1) = 2$

**(d)** There are two $x$ values leading to $f(x) = 1$, namely $x = 2$ and $x = 4$. So $f(2) = 1$ and $f(4) = 1$.

**13.** Since $f(0) = f(4) = f(8) = 0$, the solutions are $x = 0, 4, 8$.

**17. (a)** Yes. For each value of $s$, there is exactly one area.

**(b)** No. Suppose $s = 4$ represents the length of the rectangle. The width could have any other value, say 7 or 1.5 or $\pi$ or .... In this case, for one value of $s$, there are infinitely many possible values for $A$, so the area of a rectangle is not a function of the length of one of its sides.

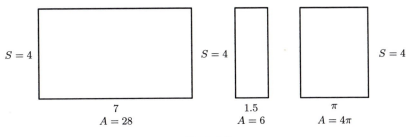

**Figure 1.2**

## Problems

**21. (a)** The number of people who own cell phones in the year 2000 is 100,300,000.
  **(b)** There are 20,000,000 people who own cell phones $a$ years after 1990.
  **(c)** There will be $b$ million people who own cell phones in the year 2010.
  **(d)** The number $n$ is the number of people (in millions) who own cell phones $t$ years after 1990.

**25. (a)** At 40 mph, fuel consumption is about 28 mpg, so the fuel used is $300/28 = 10.71$ gallons.
  **(b)** At 60 mph, fuel consumption is about 29 mpg. At 70 mph, fuel consumption is about 28 mpg. Therefore, on a 200 mile trip

$$\text{Fuel saved} = \frac{200}{28} - \frac{200}{29} = 0.25 \text{ gallons.}$$

  **(c)** The most fuel-efficient speed is where mpg is a maximum, which is about 55 mph.

**29. (a)** $69°$F
  **(b)** July $17^{\text{th}}$ and $20^{\text{th}}$
  **(c)** Yes. For each date, there is exactly one low temperature.
  **(d)** No, it is not true that for each low temperature, there is exactly one date: for example, $73°$ corresponds to both the $17^{\text{th}}$ and $20^{\text{th}}$.

**33. (a)** Adding the male total to the female total gives $x + y$, the total number of applicants.
  **(b)** Of the men who apply, 15% are accepted. So $0.15x$ male applicants are accepted. Likewise, 18% of the women are accepted so we have $0.18y$ women accepted. Summing the two tells us that $0.15x + 0.18y$ applicants are accepted.
  **(c)** The number accepted divided by the number who applied times 100 gives the percentage accepted. This expression is

$$\frac{(0.15)x + (0.18)y}{x + y}(100), \quad \text{or} \quad \frac{15x + 18y}{x + y}.$$

**37. (a)**

**Table 1.1**  *Relationship between cost, $C$, and number of liters produced, $l$*

| $l$ (millions of liters) | 0 | 1 | 2 | 3 | 4 | 5 |
|---|---|---|---|---|---|---|
| $C$ (millions of dollars) | 2.0 | 2.5 | 3.0 | 3.5 | 4.0 | 4.5 |

  **(b)** The cost, $C$, consists of a fixed cost of \$2 million plus a variable cost of \$0.50 million per million liters produced. If $l$ millions of liters are produced, the total variable costs are $(0.5)l$. Thus, the total cost $C$ in millions of dollars is given by

$$C = \text{Fixed cost} + \text{Variable cost,}$$

so

$$C = 2 + (0.5)l.$$

# Solutions for Section 1.2

## Skill Refresher

**S1.** $\frac{4-6}{3-2} = \frac{-2}{1} = -2.$

**S5.** $\frac{\frac{1}{2}-(-4)^2-\left(\frac{1}{2}-(5^2)\right)}{-4-5} = \frac{\frac{1}{2}-16-\frac{1}{2}+25}{-9} = \frac{9}{-9} = -1.$

**S9.** $\frac{x^2-\frac{3}{4}-\left(y^2-\frac{3}{4}\right)}{x-y} = \frac{x^2-\frac{3}{4}-y^2+\frac{3}{4}}{x-y} = \frac{x^2-y^2}{x-y} = \frac{(x-y)(x+y)}{x-y} = x+y.$

## Exercises

1. We take the change in number, $n$, of CDs divided by the change in time, $t$ years, to determine the average rate of change.

   (a) $\dfrac{\Delta n}{\Delta t} = \dfrac{120 - 40}{2008 - 2005} = \dfrac{80}{3}$, so the average rate of change is 80/3 CDs per year.

   (b) $\dfrac{\Delta n}{\Delta t} = \dfrac{40 - 120}{2012 - 2008} = \dfrac{-80}{4} = -20$, so the average rate of change is $-20$ CDs per year. That is, you have on average 20 fewer CDs per year.

   (c) $\dfrac{\Delta n}{\Delta t} = \dfrac{40 - 40}{2012 - 2005} = \dfrac{0}{7} = 0$, so the average rate of change is zero CDs per year.

5. We have

$$\begin{aligned}
\frac{\Delta f}{\Delta x} &= \frac{f(6.1) - f(2.2)}{6.1 - 2.2} \\
&= \frac{4.9 - 2.9}{3.9} \\
&= 0.513.
\end{aligned}$$

9. (a) Negative
   (b) Positive

## Problems

13. (a) The coordinates of point $A$ are $(10, 30)$.
       The coordinates of point $B$ are $(30, 40)$.
       The coordinates of point $C$ are $(50, 90)$.
       The coordinates of point $D$ are $(60, 40)$.
       The coordinates of point $E$ are $(90, 40)$.
    (b) From Figure 1.3, we see that $F$ is on the graph but $G$ is not.

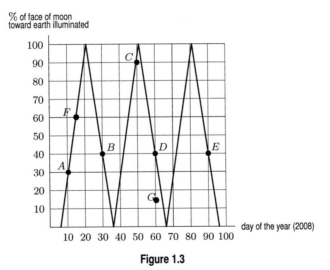

**Figure 1.3**

    (c) The function is increasing from, approximately, days 6 through 21, 36 through 51, and 66 through 81.
    (d) The function is decreasing from, approximately, days 22 through 35, 52 through 65, and 82 through 96.

**17. (a)** We see from the figure that at time $t = 0$ the population of Town A is 5000 people, whereas the population of Town B is 10,000 people. Thus, Town B starts with more people than Town A.

**(b)** Town A grows from a population of 5000 at time $t = 0$ to a population of 25,000 at $t = 50$. Thus, it grows by $25,000 - 5,000 = 20,000$ people during these 50 years, and the

$$
\begin{aligned}
\frac{\text{average rate of change}}{\text{of population A}} &= \frac{\text{change in value of } P}{\text{change in value of } t} \\
&= \frac{20,000 \text{ people}}{50 \text{ years}} \\
&= 400 \text{ people/year.}
\end{aligned}
$$

On the other hand, Town B grows from a population of 10,000 at time $t = 0$ to a population of 20,000 at time $t = 50$. Thus, it grows by $20,000 - 10,000 = 10,000$ people during these 50 years, and the

$$
\begin{aligned}
\frac{\text{average rate of change}}{\text{of population B}} &= \frac{\text{change in value of } P}{\text{change in value of } t} \\
&= \frac{10,000 \text{ people}}{50 \text{ years}} \\
&= 200 \text{ people/year.}
\end{aligned}
$$

Thus, during the time interval shown, Town A is growing twice as fast as Town B.

**21. (a)** According to the table, a 200-lb person uses 5.4 calories per minute while walking. Since a half hour is 30 minutes, a half-hour walk uses $(5.4)(30) = 162$ calories.

**(b)** A 120-lb swimmer uses 6.9 calories per minute. Thus, in one hour the swimmer uses $(6.9)(60) = 414$ calories. A 220-lb bicyclist uses 11.9 calories per minute. In a half-hour, the bicyclist uses $(11.9)(30) = 357$ calories. Thus, the swimmer uses more calories.

**(c)** Increases, since the numbers 2.7, 3.2, 4.0, 4.6, 5.4, 5.9 are increasing.

**25. (a)** Since $\Delta t$ refers to the change in the numbers of years, we calculate

$$
\Delta t = 1970 - 1960 = 10, \qquad \Delta t = 1980 - 1970 = 10, \qquad \text{and so on until 2000.}
$$

So from 1960 to 2000 $\Delta t = 10$ for all consecutive entries. From 2000 to 2007 $\Delta t = 7$, and from 2007 to 2008, $\Delta t = 1$.

**(b)** Since $\Delta G$ is the change in the amount of garbage produced per year, for the period 1960-1970 we have

$$
\Delta G = 121.1 - 88.1 = 30.
$$

Continuing in this way gives the Table 1.2:

**Table 1.2**

| Time period | 1960–70 | 1970–80 | 1980–90 | 1990–2000 | 2000–2007 | 2007–2008 |
|---|---|---|---|---|---|---|
| $\Delta G$ | 30 | 30.5 | 53.6 | 33.9 | 115.5 | $-5$ |

**(c)** Not all of the $\Delta G$ values are the same. We also know that not all the values of $\Delta t$ are the same. Computing $\Delta G / \Delta t$, we see the average rate of change in the amount of garbage produced each year, is not constant. This tells us that the amount of garbage being produced each year is changing, but not at a constant rate.

**(d)** In 2007 the United States embarked on a recycling and composting program.

# Solutions for Section 1.3

## Skill Refresher

**S1.** We have $f(0) = \frac{2}{3}(0) + 5 = 5$ and $f(3) = \frac{2}{3}(3) + 5 = 2 + 5 = 7$.

**S5.** To find the $y$-intercept, we let $x = 0$,

$$y = -4(0) + 3$$
$$= 3.$$

To find the $x$-intercept, we let $y = 0$,

$$0 = -4x + 3$$
$$4x = 3$$
$$x = \frac{3}{4}.$$

**S9.** Combining like terms we get

$$(a - 3)x - ab + a + 3.$$

Hence the constant term is $-ab + a + 3$ and the coefficient is $a - 3$.

## Exercises

**1.** The function $g$ is not linear even though $g(x)$ increases by $\Delta g(x) = 50$ each time. This is because the value of $x$ does not increase by the same amount each time. The value of $x$ increases from 0 to 100 to 300 to 600 taking steps that get larger each time.

**5.** This table could represent a linear function because the rate of change of $p(\gamma)$ is constant. Between consecutive data points, $\Delta \gamma = -1$ and $\Delta p(\gamma) = 10$. Thus, the rate of change is $\Delta p(\gamma)/\Delta \gamma = -10$. Since this is constant, the function could be linear.

**9.** The vertical intercept is 54.25, which tells us that in 1970 ($t = 0$) the population was $54,250$ (54.25 thousand) people. The slope is $-\frac{2}{7}$. Since

$$\text{Slope} = \frac{\Delta \text{population}}{\Delta \text{years}} = -\frac{2}{7},$$

we know that every seven years the population decreases by 2000 people. That is, the population decreases by 2/7 thousand per year.

## Problems

**13. (a)** To find out if this function could be linear, we calculate rates of change between pairs of points and determine whether they are constant.

$$\frac{100 - 75}{65 - 60} = \frac{25}{5} = 5 \qquad \frac{125 - 100}{70 - 65} = \frac{25}{5} = 5$$

If we continue this process, we note that the rate of change is always 5, so the function could be linear.

**(b)** Using units while calculating the rate of change, we get

$$\frac{\$100 - \$75}{65 \text{ mph} - 60 \text{ mph}} = \frac{\$25}{5 \text{ mph}} = \frac{\$5}{1 \text{ mph}}.$$

This suggests that for each increase of 1 mile per hour of speed, the fine is increased by five dollars.

**(c)** See Figure 1.4.

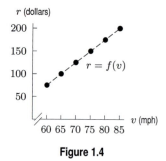

**Figure 1.4**

**17. (a)** If the relationship is linear we must show that the rate of change between any two points is the same. That is, for any two points $(x_0, C_0)$ and $(x_1, C_1)$, the quotient

$$\frac{C_1 - C_0}{x_1 - x_0}$$

is constant. From Table 1.25 we have taken the data $(0, 50)$, $(10, 52.50)$; $(5, 51.25)$, $(100, 75.00)$; and $(50, 62.50)$, $(200, 100.00)$.

$$\frac{52.50 - 50.00}{10 - 0} = \frac{2.50}{10} = 0.25$$

$$\frac{75.00 - 51.25}{100 - 5} = \frac{23.75}{95} = 0.25$$

$$\frac{100.00 - 62.50}{200 - 50} = \frac{37.50}{150} = 0.25$$

You can verify that choosing any one other pair of data points will give a slope of 0.25. The data are linear.
**(b)** The data from Table 1.25 are plotted below.

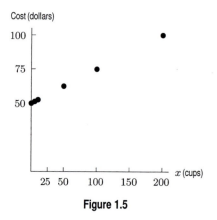

**Figure 1.5**

**(c)** Place a ruler on these points. You will see that they appear to lie on a straight line. The slope of the line equals the rate of change of the function, which is 0.25. Using units, we note that

$$\frac{\$52.50 - \$50.00}{10 \text{ cups} - 0 \text{ cups}} = \frac{\$2.50}{10 \text{ cups}} = \frac{\$0.25}{\text{cup}}.$$

In other words, the price for each additional cup of coffee is \$0.25.
**(d)** The vendor has fixed start-up costs for this venture, i.e. cart rental, insurance, salary, etc.

**21.** Each month, regardless of the amount of rocks mined from the quarry, the owners must pay 1000 dollars for maintenance and insurance, as well as 3000 dollars for monthly salaries. This totals to 4000 dollars in fixed costs. In addition, the cost for mining each ton of rocks is 80 dollars. The total cost incurred by the quarry's owners each month can be written:

$$\begin{pmatrix} \text{total} \\ \text{cost} \end{pmatrix} = \begin{pmatrix} \text{fixed} \\ \text{costs} \end{pmatrix} + \begin{pmatrix} \text{mining cost} \\ \text{per ton} \end{pmatrix} \cdot \begin{pmatrix} \text{tons of rocks} \\ \text{mined} \end{pmatrix}$$

$$= 4000 \text{ dollars} + (80 \text{ dollars/ton})(r \text{ tons})$$

$$c = 4000 + 80r.$$

**25. (a)** No. The values of $f(d)$ first drop, then rise, so $f$ is not linear.
   **(b)** For $d \geq 150$, the graph looks linear. See Figure 1.6.

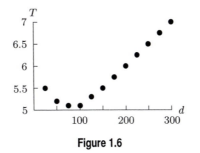

**Figure 1.6**

   **(c)** For $d \geq 150$ the average rate of change appears to be constant. Each time the depth goes up by $\Delta d = 25$ meters, the temperature rises by $\Delta T = 0.25°$C, so the average rate of change is $\Delta T / \Delta d = 0.25/25 = 0.01°$C/meter. In other words, the temperature rises by $0.01°$C for each extra meter in depth.

**29. (a)** Here we have

$$2r = 5$$
$$r = \frac{5}{2}$$
$$\text{and} \quad \sqrt{s} = 4$$
$$s = 16$$
$$\text{giving} \quad y = 2 \underbrace{\left(\frac{5}{2}\right)}_{2r} + x \underbrace{\sqrt{16}}_{\sqrt{s}}.$$

   **(b)** Here we have

$$\frac{1}{k} = 5$$
$$k = \frac{1}{5} = 0.2$$
$$\text{and} \quad -(j-1) = 4$$
$$j = -3$$
$$\text{giving}$$
$$y = \underbrace{\frac{1}{0.2}}_{1/k} - \underbrace{(-3-1)}_{j-1} x.$$

**33.** Most functions look linear if viewed in a small enough window. This function is not linear. We see this by graphing the function in the larger window $-100 \leq x \leq 100, -20 \leq y \leq 20$.

## Solutions for Section 1.4

### Skill Refresher

**S1.**
$$y - 5 = 21$$
$$y = 26.$$

**S5.** We first distribute $\frac{5}{3}(y + 2)$ to obtain:

$$\frac{5}{3}(y + 2) = \frac{1}{2} - y$$
$$\frac{5}{3}y + \frac{10}{3} = \frac{1}{2} - y$$
$$\frac{5}{3}y + y = \frac{1}{2} - \frac{10}{3}$$
$$\frac{5}{3}y + \frac{3y}{3} = \frac{3}{6} - \frac{20}{6}$$
$$\frac{8y}{3} = -\frac{17}{6}$$
$$\left(\frac{3}{8}\right)\frac{8y}{3} = \left(\frac{3}{8}\right)\left(-\frac{17}{6}\right)$$
$$y = -\frac{17}{16}.$$

**S9.** We collect all terms involving $x$ and then divide by $2a$:

$$ab + ax = c - ax$$
$$2ax = c - ab$$
$$x = \frac{c - ab}{2a}.$$

### Exercises

**1.** Rewriting in slope-intercept form:

$$5(x + y) = 4$$
$$5x + 5y = 4$$
$$5y = 4 - 5x$$
$$\frac{5y}{5} = \frac{4}{5} - \frac{5x}{5}$$
$$y = \frac{4}{5} - x$$

**5.** Rewriting in slope-intercept form:

$$y - 0.7 = 5(x - 0.2)$$
$$y - 0.7 = 5x - 1$$
$$y = 5x - 1 + 0.7$$
$$y = 5x - 0.3$$
$$y = -0.3 + 5x$$

**9.** Rewriting in slope-intercept form:

$$\frac{x+y}{7} = 3$$
$$x + y = 21$$
$$y = 21 - x$$

**13.** Yes. Write the function as

$$C(r) = 0 + 2\pi r,$$

so $C(r)$ is linear with $b = 0$ and $m = 2\pi$.

**17.** We can put the slope $m = 3$ and $y$-intercept $b = 8$ directly into the general equation $y = b + mx$ to get $y = 8 + 3x$.

**21.** Since the slope is $m = 0.1$, the equation is

$$y = b + 0.1x.$$

Substituting $x = -0.1$, $y = 0.02$ to find $b$ gives

$$0.02 = b + 0.1(-0.1)$$
$$0.02 = b - 0.01$$
$$b = 0.03.$$

The equation is $y = 0.03 + 0.1x$.

**25.** Since the function is linear, we can choose any two points to find its formula. We use the form

$$q = b + mp$$

to get the number of bottles sold as a function of the price per bottle. We use the two points $(0.50, 1500)$ and $(1.00, 500)$. We begin by finding the slope, $\Delta q/\Delta p = (500 - 1500)/(1.00 - 0.50) = -2000$. Next, we substitute a point into our equation using our slope of $-2000$ bottles sold per dollar increase in price and solve to find $b$, the $q$-intercept. We use the point $(1.00, 500)$:

$$500 = b - 2000 \cdot 1.00$$
$$2500 = b.$$

Therefore,

$$q = 2500 - 2000p.$$

**29.** Since the function is linear, we can use any two points (from the graph) to find its formula. We use the form

$$u = b + mn$$

to get the meters of shelf space used as a function of the number of different medicines stocked. We use the two points $(60, 5)$ and $(120, 10)$. We begin by finding the slope, $\Delta u/\Delta n = (10 - 5)/(120 - 60) = 1/12$. Next, we substitute a point into our equation using our slope of $1/12$ meters of shelf space per medicine and solve to find $b$, the $u$-intercept. We use the point $(60, 5)$:

$$5 = b + (1/12) \cdot 60$$
$$0 = b.$$

Therefore,

$$u = (1/12)n.$$

The fact that $b = 0$ is not surprising, since we would expect that, if no medicines are stocked, they should take up no shelf space.

## Problems

**33.** The starting value is $b = 12{,}000$, and the growth rate is $m = 225$, so $h(t) = 12{,}000 + 225t$.

**37.** We found in Problem 36 that the fixed cost of this good is \$10,500. We found the unit cost in Problem 35(c) to be \$5. (In that problem, we used $n = 150$ and $n = 175$, but since this is a linear total-cost function, any pair of values of $n$ will give the same rate of change or cost per unit.) Thus,

$$C(n) = 10{,}500 + 5n.$$

**41. (a)** Since $q$ is linear, $q = b + mp$, where

$$m = \frac{\Delta q}{\Delta p} = \frac{65 - 45}{3.10 - 3.50}$$
$$= \frac{20}{-0.40} = -50 \text{ gallons/dollar.}$$

Thus, $q = b - 50p$ and since $q = 65$ if $p = 3.10$,

$$65 = b - 50(3.10)$$
$$65 = b - 155$$
$$b = 65 + 155 = 210.$$

So,

$$q = 210 - 50p.$$

**(b)** The slope is $m = -50$ gallons per dollar, which tells us that the quantity of gasoline demanded in one time period decreases by 50 gallons for each \$1 increase in price.

**(c)** If $p = 0$ then $q = 210$, which means that if the price of gas were \$0 per gallon, then the quantity demanded in one time period would be 210 gallons per month. This means if gas were free, a person would want 210 gallons. If $q = 0$ then $210 - 50p = 0$, so $210 = 50p$ and $p = 210/50 = 4.20$. This tells us that (according to the model), at a price of \$4.20 per gallon there will be no demand for gasoline. In the real world, this is not likely.

**45.** Both $P$ and $Q$ lie on $y = x^2 + 1$, so their coordinates must satisfy that equation. Point $Q$ has $x$-coordinate 2, so $y = 2^2 + 1 = 5$. Point $P$ has $y$-coordinate 8, so

$$8 = x^2 + 1$$
$$x^2 = 7,$$

and $x = -\sqrt{7}$ because we know from the graph that $x < 0$.

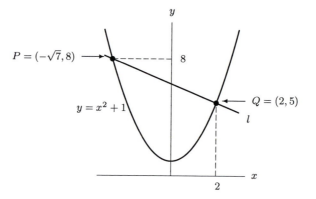

**Figure 1.7**

Using the coordinates of $P$ and $Q$, we know that

$$m = \frac{\Delta y}{\Delta x} = \frac{5 - 8}{2 - (-\sqrt{7})} = \frac{-3}{2 + \sqrt{7}}.$$

Since $y = 5$ when $x = 2$, we have

$$5 = b - \frac{3}{2 + \sqrt{7}} \cdot 2$$

$$5 = b - \frac{6}{2 + \sqrt{7}}$$

$$b = 5 + \frac{6}{2 + \sqrt{7}} = \frac{5(2 + \sqrt{7})}{(2 + \sqrt{7})} + \frac{6}{2 + \sqrt{7}}$$

$$= \frac{10 + 5\sqrt{7} + 6}{2 + \sqrt{7}} = \frac{16 + 5\sqrt{7}}{2 + \sqrt{7}}.$$

So the equation of the line is

$$y = \frac{16 + 5\sqrt{7}}{2 + \sqrt{7}} - \frac{3}{2 + \sqrt{7}}x.$$

Optionally, this can be simplified (by rationalizing denominators) to

$$y = 1 + 2\sqrt{7} + (2 - \sqrt{7})x.$$

**49. (a)** Since $i$ is linear, we can write

$$i(x) = b + mx.$$

Since $i(10) = 25$ and $i(20) = 50$, we have

$$m = \frac{50 - 25}{20 - 10} = 2.5.$$

So,

$$i(x) = b + 2.5x.$$

Using $i(10) = 25$, we can solve for $b$:

$$i(10) = b + 2.5(10)$$
$$25 = b + 25$$
$$b = 0.$$

Our formula then is

$$i(x) = 2.5x.$$

**(b)** The increase in risk associated with *not* smoking is $i(0)$. Since there is no increase in risk for a non-smoker, we have $i(0) = 0$.

**(c)** The slope of $i(x)$ tells us that the risk increases by a factor of 2.5 with each additional cigarette a person smokes per day.

**53. (a)** We know that $r = 1/t$. Table 1.3 gives values of $r$. From the table, we see that $\Delta r / \Delta H \approx 0.01/2 = 0.005$, so $r = b + 0.005H$. Solving for $b$, we have

$$0.070 = b + 0.005 \cdot 20$$
$$b = 0.070 - 0.1 = -0.03.$$

Thus, a formula for $r$ is given by $r = 0.005H - 0.03$.

**Table 1.3**  *Development time t (in days) for an organism as a function of ambient temperature H (in °C)*

| $H$, °C  | 20    | 22    | 24    | 26    | 28    | 30    |
|----------|-------|-------|-------|-------|-------|-------|
| $r$, rate | 0.070 | 0.080 | 0.090 | 0.100 | 0.110 | 0.120 |

   **(b)** From Problem 52, we know that if $r = b + kH$ then the number of degree-days is given by $S = 1/k$. From part (a) of this problem, we have $k = 0.005$, so $S = 1/0.005 = 200$.

# Solutions for Section 1.5

## Skill Refresher

**S1.** Plugging 5 in the first equation for $y$ we get

$$x + 5 = 3$$
$$x = -2.$$

**S5.** Substituting the value of $y$ from the first equation into the second equation, we obtain

$$x + 2(2x - 10) = 15$$
$$x + 4x - 20 = 15$$
$$5x = 35$$
$$x = 7.$$

Now we substitute $x = 7$ into the first equation, obtaining $2(7) - y = 10$, hence $y = 4$.

## Exercises

1. **(a)** is (V), because slope is negative, vertical intercept is 0
   **(b)** is (VI), because slope and vertical intercept are both positive
   **(c)** is (I), because slope is negative, vertical intercept is positive
   **(d)** is (IV), because slope is positive, vertical intercept is negative
   **(e)** is (III), because slope and vertical intercept are both negative
   **(f)** is (II), because slope is positive, vertical intercept is 0

5. **(a)** See Figures 1.8 and 1.9.

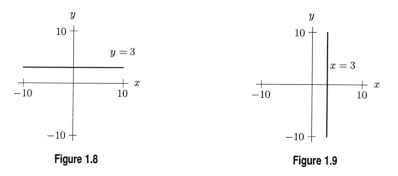

Figure 1.8                    Figure 1.9

   **(b)** Yes for $y = 3$: $y = 3 + 0x$. No for $x = 3$, since the slope is undefined, and there is no $y$-intercept.

9. These lines are neither parallel nor perpendicular. They do not have the same slope, nor are their slopes negative reciprocals (if they were, one of the slopes would be negative).

## Problems

13. We can write the equation in slope-intercept form

$$3x + 5y = 6$$
$$5y = 6 - 3x$$
$$y = \frac{6}{5} - \frac{3}{5}x.$$

The slope is $\frac{-3}{5}$. Lines parallel to this line all have slope $\frac{-3}{5}$. Since the line passes through $(0, 6)$, its $y$-intercept is equal to 6. So $y = 6 - \frac{3}{5}x$.

17. Since $P$ is the $x$-intercept, we know that point $P$ has $y$-coordinate $= 0$, and if the $x$-coordinate is $x_0$, we can calculate the slope of line $l$ using $P(x_0, 0)$ and the other given point $(0, -2)$.

$$m = \frac{-2 - 0}{0 - x_0} = \frac{-2}{-x_0} = \frac{2}{x_0}.$$

We know this equals 2, since $l$ is parallel to $y = 2x + 1$ and therefore must have the same slope. Thus we have

$$\frac{2}{x_0} = 2.$$

So $x_0 = 1$ and the coordinates of $P$ are $(1, 0)$.

21. (a) This line, being parallel to $l$, has the same slope. Since the slope of $l$ is $-\frac{2}{3}$, the equation of this line is

$$y = b - \frac{2}{3}x.$$

To find $b$, we use the fact that $P = (6, 5)$ is on this line. This gives

$$5 = b - \frac{2}{3}(6)$$
$$5 = b - 4$$
$$b = 9.$$

So the equation of the line is

$$y = 9 - \frac{2}{3}x.$$

(b) This line is perpendicular to line $l$, and so its slope is given by

$$m = \frac{-1}{-2/3} = \frac{3}{2}.$$

Therefore its equation is

$$y = b + \frac{3}{2}x.$$

We again use point P to find $b$:

$$5 = b + \frac{3}{2}(6)$$
$$5 = b + 9$$
$$b = -4.$$

This gives

$$y = -4 + \frac{3}{2}x.$$

(c) Figure 1.10 gives a graph of line $l$ together with point $P$ and the two lines we have found.

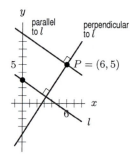

**Figure 1.10**: Line $l$ and two lines through $P$, one parallel and one perpendicular to $l$

25. The graphs are shown in Figure 1.11.

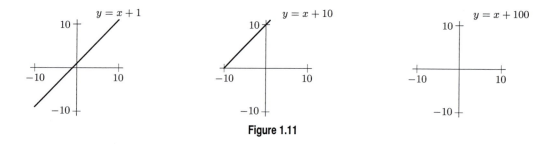

**Figure 1.11**

(a) As $b$ becomes larger, the graph moves higher and higher up, until it disappears from the viewing rectangle.
(b) There are many correct answers, one of which is $y = x - 100$.

29. (a) Use the point-slope formula:
$$y - 0 = \frac{\sqrt{3}}{-1}(x - 0),$$
so $y = -\sqrt{3}x$.

(b) The slope of the tangent line is the negative reciprocal of $-\sqrt{3}$, so $m = \dfrac{1}{\sqrt{3}}$, and
$$y - \sqrt{3} = \frac{1}{\sqrt{3}}(x - (-1)),$$
or
$$y = \frac{1}{\sqrt{3}}x + \frac{4}{\sqrt{3}}.$$

## Solutions for Section 1.6

1. Although the points do not lie on a line, they are tending upward as $x$ increases. So, there is a strong positive correlation; $r = 0.93$ is reasonable.

**5.** A scatter plot of the data is shown in Figure 1.12.

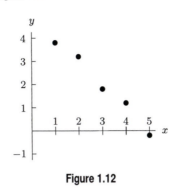

**Figure 1.12**

$r = 1$ is not possible. These points are very close to lying on a line with negative slope, so $r$ will be negative. $(r = -0.98.)$

**9. (a)** See Figure 1.13.

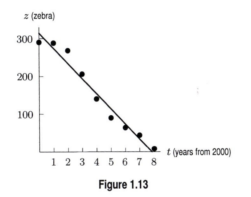

**Figure 1.13**

**(b)** Estimates vary, but should be roughly $z = -40t + 314$.

**(c)** A calculator gives $z = -40t + 314$.

**(d)** The slope of $-40$ tells us that, on average, 40 zebra die per year. The vertical intercept value is the initial population. The vertical intercept of the line is 314, which is close to the initial data value of 290. The horizontal intercept of 7.85 years is the number of years until all the zebra have died.

**(e)** There is a strong negative correlation, $(r = -0.983)$.

# Solutions for Chapter 1 Review

## Exercises

**1.** Any $w$ value can give two $z$ values. For example, if $w = 0$,

$$7 \cdot 0^2 + 5 = z^2$$
$$\pm\sqrt{5} = z,$$

so there are two $z$ values (one positive and one negative) for $w = 0$. Thus, $z$ is not a function of $w$.

A similar argument shows that $w$ is not a function of $z$.

**5.** At the two points where the graph breaks (marked $A$ and $B$ in Figure 1.14), there are two $y$ values for a single $x$ value. The graph does not pass the vertical line test. Thus, $y$ is not a function of $x$.

Similarly, $x$ is not a function of $y$ because there are many $y$ values that give two $x$ values (For example, $y = 0$.)

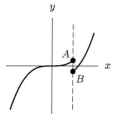

**Figure 1.14**

**9. (a)** Machine #2 gives two different possible snacks for each button. Thus, $S$ is not a function of $N$. It is a bad machine to use because you can't choose the snack you will get.

**(b)** Machines #1 and #3 give $S$ as a function of $N$. This means that by choosing a button number, you can choose a snack.

**(c)** Machine #3. $N$ is not a function of $S$ because two different button numbers correspond to the same snack. For example, $N = 8$ and 9 both correspond to Snickers. This means that one snack corresponds to more than one button.

**13.** This table could represent a linear function, because, for the values shown, the rate of change of $a(t)$ is constant. For the given data points, between consecutive points, $\Delta t = 3$, and $\Delta a(t) = 2$. Thus, in each case, the rate of change is $\Delta a(t)/\Delta t = 2/3$. Since the rate of change is constant, the function could be linear.

**17. (a)** $f(x)$ has a $y$-intercept of 1 and a positive slope. Thus, (ii) must be the graph of $f(x)$.

**(b)** $g(x)$ has a $y$-intercept of 1 and a negative slope. Thus, (iii) must be the graph of $g(x)$.

**(c)** $h(x)$ is a constant function with a $y$ intercept of 1. Thus, (i) must be the graph of $h(x)$.

**21.** These lines are neither parallel nor perpendicular. They do not have the same slope, nor are their slopes negative reciprocals (if they were, one of the slopes would be negative).

## Problems

**25.** From the table, $r(300) = 120$, which tells us that at a height of 300 m the wind speed is 120 mph.

**29.** A possible graph is shown in Figure 1.15.

distance of bug from light

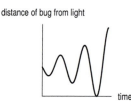

time

**Figure 1.15**

**33.** The diagram is shown in Figure 1.16.

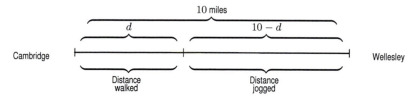

**Figure 1.16**

The total time the trip takes is given by the equation

$$\text{Total time} = \text{Time walked} + \text{Time jogged}.$$

The distance walked is $d$, and, since the total distance is 10, the remaining distance jogged is $(10 - d)$. See Figure 1.16. We know that time equals distance over speed, which means that

$$\text{Time walked} = \frac{d}{5} \quad \text{and} \quad \text{Time jogged} = \frac{10 - d}{8}.$$

Thus, the total time is given by the equation

$$T(d) = \frac{d}{5} + \frac{10 - d}{8}.$$

**37. (a)** (i) Between $(-1, f(-1))$ and $(3, f(3))$

$$\text{Average rate of change} = \frac{f(3) - f(-1)}{3 - (-1)} = \frac{\left(\frac{3}{2} + \frac{5}{2}\right) - \left(\frac{-1}{2} + \frac{5}{2}\right)}{4} = \frac{4 - 2}{4} = \frac{2}{4} = \frac{1}{2}.$$

(ii) Between $(a, f(a))$ and $(b, f(b))$

$$\text{Average rate of change} = \frac{f(b) - f(a)}{b - a} = \frac{\left(\frac{b}{2} + \frac{5}{2}\right) - \left(\frac{a}{2} + \frac{5}{2}\right)}{b - a} = \frac{\frac{b}{2} + \frac{5}{2} - \frac{a}{2} - \frac{5}{2}}{b - a} = \frac{\frac{b}{2} - \frac{a}{2}}{b - a} = \frac{\frac{1}{2}(b - a)}{b - a} = \frac{1}{2}.$$

(iii) Between $(x, f(x))$ and $(x + h, f(x + h))$

$$\text{Average rate of change} = \frac{f(x + h) - f(x)}{(x + h) - x} = \frac{\left(\frac{x+h}{2} + \frac{5}{2}\right) - \left(\frac{x}{2} + \frac{5}{2}\right)}{(x + h) - x}$$

$$= \frac{\frac{x+h}{2} + \frac{5}{2} - \frac{x}{2} - \frac{5}{2}}{x + h - x} = \frac{\frac{x+h-x}{2}}{h} = \frac{\frac{h}{2}}{h} = \frac{1}{2}.$$

**(b)** The average rate of change is always $\frac{1}{2}$.

**41. (a)** One horse costs the woodworker \$5000 in start-up costs plus \$350 for labor and materials, a total of \$5350. Thus, if $n = 1$, we have

$$C = \underbrace{5000}_{\text{Start-up costs}} + \underbrace{350}_{\text{Extra cost for 1 horse}} = 5350.$$

Similarly, for 2 horses

$$C = \underbrace{5000}_{\text{Start-up costs}} + \underbrace{350 \cdot 2}_{\text{Extra cost for 2 horses}} = 5700,$$

and for 5 horses

$$C = \underbrace{5000}_{\text{Start-up costs}} + \underbrace{350 \cdot 5}_{\text{Extra cost for 5 horses}} = 6750.$$

Similarly, for 10 horses, $C = 5000 + 350 \cdot 10 = 8500$, and for 20 horses, $C = 12{,}000$. See Table 1.4 and Figure 1.17.

**Table 1.4**  *Total cost of carving n horses*

| $n$, number of horses | $C$, total cost (\$) |
|---|---|
| 0 | 5000 |
| 1 | 5350 |
| 2 | 5700 |
| 5 | 6750 |
| 10 | 8500 |
| 20 | 12,000 |

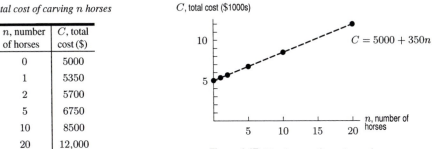

**Figure 1.17**: Total cost of carving $n$ horses

Notice that it costs the woodworker $5000 to carve 0 horses since he buys the tools, plans, and advertising even if he never carves a single horse.

**(b)** From part (a), a formula for $C$, as a function of $n$, is

$$C = \underbrace{5000}_{\text{Start-up costs}} + \underbrace{350 \cdot n}_{\text{Extra cost for } n \text{ horses}} = 5000 + 350n.$$

**(c)** The average rate of change of this function is

$$\text{Rate of change} = \frac{\Delta C}{\Delta n} = \frac{\text{Change in cost}}{\text{Change in number of horses carved}}.$$

Each additional horse costs an extra $350, so

$$\Delta C = 350 \quad \text{if} \quad \Delta n = 1.$$

Thus, the rate of change is given by

$$\frac{\Delta C}{\Delta n} = \frac{\$350}{1 \text{ horse}} = \$350 \text{ per horse.}$$

The rate of change of $C$ gives the additional cost to carve one additional horse. Since the total cost increases at a constant rate ($350 per horse), the graph of $C$ against $n$ is a straight line sloping upward.

**45. (a)** The results are in Table 1.5.

**Table 1.5**

| $t$ | 0 | 1 | 2 | 3 | 4 |
|---|---|---|---|---|---|
| $v = f(t)$ | 1000 | 990.2 | 980.4 | 970.6 | 960.8 |

**(b)** The speed of the bullet is decreasing at a constant rate of 9.8 meters/sec every second. To confirm this, calculate the rate of change in velocity over every second. We get

$$\frac{\Delta v}{\Delta t} = \frac{990.2 - 1000}{1 - 0} = \frac{980.4 - 990.2}{2 - 1} = \frac{970.6 - 980.4}{3 - 2} = \frac{960.8 - 970.6}{4 - 3} = -9.8.$$

Since the value of $\Delta v/\Delta t$ comes out the same, $-9.8$, for every interval, we can say that the bullet is slowing down at a constant rate. This makes sense as the constant force of gravity acts to slow the upward moving bullet down at a constant rate.

**(c)** The slope, $-9.8$, is the rate at which the velocity is changing. The $v$-intercept of 1000 is the initial velocity of the bullet. The $t$-intercept of $1000/9.8 = 102.04$ is the time at which the bullet stops moving and starts to head back to Earth.

**(d)** Since Jupiter's gravitational field would exert a greater pull on the bullet, we would expect the bullet to slow down at a faster rate than a bullet shot from earth. On earth, the rate of change of the bullet is $-9.8$, meaning that the bullet is slowing down at the rate of 9.8 meters per second. On Jupiter, we expect that the coefficient of $t$, which represents the rate of change, to be a more negative number (less than $-9.8$). Similarly, since the gravitational pull near the surface of the moon is less, we expect that the bullet would slow down at a lesser rate than on earth. So, the coefficient of $t$ should be a less negative number (greater than $-9.8$ but less than 0).

**49. (a)** Since the relationship is linear, the general formula for $S$ in terms of $p$ is

$$S = b + mp.$$

Since we know that the quantity supplied rises by 50 units when the rise in the price is $0.50, we can write $\Delta S = 50$ units, when $\Delta p = \$0.50$. The slope is then:

$$m = \frac{\Delta S}{\Delta p} = \frac{50 \text{ units}}{\$0.50} = 100 \text{ units/dollar.}$$

Put this value of the slope into the formula for $S$ and solve for $b$ using $p = 2$ and $S = 100$:

$$S = b + mp$$
$$100 = b + (100)(2)$$
$$100 = b + 200$$
$$b = -100.$$

We now have the slope $m$ and the $S$-intercept $b$. So, we know that

$$S = -100 + 100p.$$

**(b)** The slope in this problem is 100 units/dollar, which means that for every increase of $1 in price, suppliers are willing to supply another 100 units.

**(c)** The price below which suppliers will not supply the good is represented by the point at which $S = 0$. Putting $S = 0$ into the equation found in (b) we get:

$$0 = -100 + 100p$$
$$100 = 100p$$
$$p = 1.$$

So when the price is $1, or less, the suppliers will not want to produce anything.

**(d)** From Problem 48 we know that

$$D = 1100 - 200p.$$

To find when supply equals demand set the formulas for $S$ and $D$ equal and solve for $p$:

$$S = D$$
$$-100 + 100p = 1100 - 200p$$
$$100p + 200p = 1100 + 100$$
$$300p = 1200$$
$$p = \frac{1200}{300} = 4.$$

Therefore, the market clearing price is $4.

**53.** The line intersects $f$ at $x = -2$ and $x = 5$. We see that

$$f(-2) = 2 + \frac{3}{-2+5} = 2 + \frac{3}{3} = 3$$
$$f(5) = 2 + \frac{3}{5+5} = 2 + \frac{3}{10} = 2.3,$$

so the line contains the points $(-2, 3)$ and $(5, 2.3)$. This means the slope is

$$m = \frac{2.3 - 3}{5 - (-2)} = \frac{-0.7}{7} = -0.1.$$

Using the point-slope formula with $(x_0, y_0) = (-2, 3)$, we have

$$y = 3 - 0.1\,(x - (-2))$$
$$= 2.8 - 0.1x.$$

Checking our answer, we see that

$$\text{At } x = -2: \quad y = 2.8 - 0.1(-2) = 3$$
$$\text{At } x = 5: \quad y = 2.8 - 0.1(5) \quad = 2.3,$$

as required.

**57.** First we place $5x - 3y = 6$ into slope-intercept form:

$$5x - 3y = 6$$
$$3y = -6 + 5x$$
$$y = -2 + \frac{5}{3}x.$$

The slope of this line is $5/3$. Since the graph of $g$ is perpendicular to it, the slope of $g$ is

$$m = -\frac{1}{5/3} = -\frac{3}{5}.$$

We see from the first equation that at $x = 15$,

$$y = \frac{5}{3} \cdot 15 - 2 = 25 - 2 = 23,$$

so the graph of the first equation contains the point $(15, 23)$. Since the graph of $g$ also contains this point, we have

$$g(15) = b + m \cdot 15 = 23$$
$$b - \frac{3}{5}(15) = 23$$
$$b - 9 = 23$$
$$b = 32,$$

so $g(x) = 32 - (3/5)x$.

**61. (a)** See Figure 1.18.
   **(b)** The scatterplot suggests that as IQ increases, the number of hours of TV viewing decreases. The points, though, are not close to being on a line, so a reasonable guess is $r \approx -0.5$.
   **(c)** A calculator gives the regression equation $y = 27.5139 - 0.1674x$ with $r = -0.5389$.

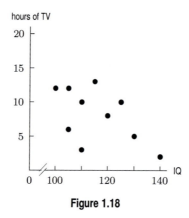

**Figure 1.18**

**65.** Writing

$$y = -3 - \frac{x}{2}$$
$$= -3 + \left(-\frac{1}{2}\right) \cdot x,$$

we see that in order for the coefficients of $x$ to be the same, we have

$$-r^2 = -\frac{1}{2}$$

$$r^2 = \frac{1}{2}$$
$$r = \sqrt{0.5}. \quad \text{because } r > 0$$

Likewise, in order for the constant terms to be the same, we have

$$\frac{p}{p-1} = -3$$
$$p = -3(p-1)$$
$$p = -3p + 3$$
$$4p = 3$$
$$p = 0.75,$$
$$\text{so we have } y = \frac{0.75}{0.75-1} - \left(\sqrt{0.5}\right)^2 x.$$

Checking our answer, we see that

$$y = \frac{0.75}{0.75-1} - \left(\sqrt{0.5}\right)^2 x.$$
$$= \frac{0.75}{-0.25} - 0.5x$$
$$= -3 - \frac{x}{2}. \qquad \text{as required}$$

**69.** In Figure 1.19 the graph of the hair length is steepest just after each haircut, assumed to be at the beginning of each year. As the year progresses, the growth is slowed by split ends. By the end of the year, the hair is breaking off as fast as it is growing, so the graph has leveled off. At this time the hair is cut again. Once again it grows until slowed by the split ends. Then it is cut. This continues for five years when the longest hairs fall out because they have come to the end of their natural lifespan.

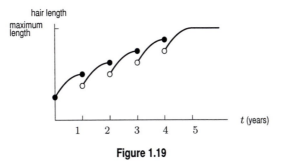

**Figure 1.19**

**73.** Extending the pattern in the table, we see that entry $n = 54$ is found in column $c = 8$, row $r = 5$. Thus, $g(54) = 8^{1/5} = \sqrt[5]{8}$.

## CHECK YOUR UNDERSTANDING

**1.** False. $f(t)$ is functional notation, meaning that $f$ is a function of the variable $t$.

**5.** True. The number of people who enter a store in a day and the total sales for the day are related, but neither quantity is uniquely determined by the other.

**9.** True. A circle does not pass the vertical line test.

**13.** True. This is the definition of an increasing function.

**17.** False. Parentheses must be inserted. The correct ratio is $\dfrac{(10 - 2^2) - (10 - 1^2)}{2 - 1} = -3$.

**21.** False. Writing the equation as $y = (-3/2)x + 7/2$ shows that the slope is $-3/2$.

**25.** True. A constant function has slope zero. Its graph is a horizontal line.

**29.** True. At $y = 0$, we have $4x = 52$, so $x = 13$. The $x$-intercept is $(13, 0)$.

**33.** False. Substitute the point's coordinates in the equation: $-3 - 4 \neq -2(4 + 3)$.

**37.** False. The first line does but the second, in slope-intercept form, is $y = (1/8)x + (1/2)$, so it crosses the $y$-axis at $y = 1/2$.

**41.** True. The point $(1, 3)$ is on both lines because $3 = -2 \cdot 1 + 5$ and $3 = 6 \cdot 1 - 3$.

**45.** True. The slope, $\Delta y / \Delta x$ is undefined because $\Delta x$ is zero for any two points on a vertical line.

**49.** False. For example, in children there is a high correlation between height and reading ability, but it is clear that neither causes the other.

**53.** True. There is a perfect fit of the line to the data.

# Solutions to Skills for Chapter 1

**1.**
$$3x = 15$$
$$\frac{3x}{3} = \frac{15}{3}$$
$$x = 5$$

**5.**
$$w - 23 = -34$$
$$w = -11$$

**9.** The common denominator for this fractional equation is 3. If we multiply both sides of the equation by 3, we obtain:
$$3\left(3t - \frac{2(t-1)}{3}\right) = 3(4)$$
$$9t - 2(t-1) = 12$$
$$9t - 2t + 2 = 12$$
$$7t + 2 = 12$$
$$7t = 10$$
$$t = \frac{10}{7}.$$

**13.** Dividing by $w$ gives $l = A/w$.

**17.** We collect all terms involving $v$ and then factor out the $v$.
$$u(v + 2) + w(v - 3) = z(v - 1)$$
$$uv + 2u + wv - 3w = zv - z$$
$$uv + wv - zv = 3w - 2u - z$$
$$v(u + w - z) = 3w - 2u - z$$
$$v = \frac{3w - 2u - z}{u + w - z}.$$

**21.** Solving for $y'$,

$$y'y^2 + 2xyy' = 4y$$
$$y'(y^2 + 2xy) = 4y$$
$$y' = \frac{4y}{y^2 + 2xy}$$
$$y' = \frac{4}{y + 2x} \text{ if } y \neq 0.$$

Note that if $y = 0$, then $y'$ could be any real number.

**25.** We substitute the expression $-\frac{3}{5}x + 6$ for $y$ in the first equation.

$$2x + 3y = 7$$
$$2x + 3\left(-\frac{3}{5}x + 6\right) = 7$$
$$2x - \frac{9}{5}x + 18 = 7 \quad \text{or}$$
$$\frac{10}{5}x - \frac{9}{5}x + 18 = 7$$
$$\frac{1}{5}x + 18 = 7$$
$$\frac{1}{5}x = -11$$
$$x = -55$$
$$y = -\frac{3}{5}(-55) + 6$$
$$y = 39$$

**29.** Substituting $y = 2x$ into $2x + y = 12$ gives

$$2x + 2x = 12$$
$$4x = 12$$
$$x = 3.$$

Thus, substituting $x = 3$ into $y = 2x$ gives $y = 6$, so the point of intersection is $x = 3$, $y = 6$.

**33.** The radius is 8, so $A = (2, 9)$, $B = (10, 1)$.

# CHAPTER TWO

## Solutions for Section 2.1

### Skill Refresher

**S1.** $5(x - 3) = 5x - 15$.

**S5.**

$$3\left(1 + \frac{1}{x}\right) = 3\left(\frac{x + 1}{x}\right)$$
$$= \frac{3x + 3}{x}.$$

**S9.**

$$\frac{21}{z - 5} - \frac{13}{z^2 - 5z} = 3$$
$$\frac{21}{z - 5} - \frac{13}{z(z - 5)} = 3$$
$$\frac{21z - 13}{z(z - 5)} = 3$$
$$21z - 13 = 3z(z - 5)$$
$$21z - 13 = 3z^2 - 15z$$
$$3z^2 - 36z + 13 = 0$$

$$z = \frac{-(-36) \pm \sqrt{(-36)^2 - 4(3)(13)}}{2(3)}$$
$$= \frac{36 \pm \sqrt{1140}}{6}$$
$$= \frac{36 \pm \sqrt{4 \cdot 285}}{6}$$
$$= \frac{36 \pm 2\sqrt{285}}{6}$$
$$= \frac{18 \pm \sqrt{285}}{3}.$$

### Exercises

1. **(a)** Substituting $t = 0$ gives $f(0) = 0^2 - 4 = -4$.
   **(b)** Setting $f(t) = 0$ and solving gives $t^2 - 4 = 0$, so $t^2 = 4$, so $t = \pm 2$.

5. Substituting $-27$ for $x$ gives

$$g(-27) = -\frac{1}{2}(-27)^{1/3} = -\frac{1}{2}(-3) = \frac{3}{2}.$$

**9.**

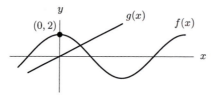

## Problems

**13.** The input, $t$, is the number of months since January 1, and the output, $F$, is the number of foxes. The expression $g(9)$ represents the number of foxes in the park on October 1. Table 1.3 on page 5 of the text gives $F = 100$ when $t = 9$. Thus, $g(9) = 100$. On October 1, there were 100 foxes in the park.

**17. (a)** (i) $\dfrac{\frac{1}{t}}{\frac{1}{t} - 1} = \dfrac{\frac{1}{t}}{\frac{1-t}{t}} = \dfrac{1}{t} \cdot \dfrac{t}{1-t} = \dfrac{1}{1-t}.$

(ii) $\dfrac{\frac{1}{t+1}}{\frac{1}{t+1} - 1} = \dfrac{1}{t+1} \cdot \dfrac{t+1}{1-t-1} = -\dfrac{1}{t}.$

**(b)** Solve $f(x) = \dfrac{x}{(x-1)} = 3$, so

$$x = 3x - 3$$
$$3 = 2x$$
$$x = \frac{3}{2}.$$

**21. (a)** The car's position after 2 hours is denoted by the expression $s(2)$. The position after 2 hours is

$$s(2) = 11(2)^2 + 2 + 100 = 44 + 2 + 100 = 146.$$

**(b)** This is the same as asking the following question: "For what $t$ is $v(t) = 65$?"
**(c)** To find out when the car is going 67 mph, we set $v(t) = 67$. We have

$$22t + 1 = 67$$
$$22t = 66$$
$$t = 3.$$

The car is going 67 mph at $t = 3$, that is, 3 hours after starting. Thus, when $t = 3$, $S(3) = 11(3^2) + 3 + 100 = 202$, so the car's position when it is going 67 mph is 202 miles.

**25.** $r(0.5s_0)$ is the wind speed at a half the height above ground of maximum wind speed.

**29.** $f(a) = \dfrac{a \cdot a}{a + a} = \dfrac{a^2}{2a} = \dfrac{a}{2}.$

**33. (a)** (i) From the table, $N(150) = 6$. When 150 students enroll, there are 6 sections.

(ii) Since $N(75) = 4$ and $N(100) = 5$, and 80 is between 75 and 100 students, we choose the higher value for $N(s)$. So $N(80) = 5$. When 80 students enroll, there are 5 sections.

(iii) The quantity $N(55.5)$ is not defined, since 55.5 is not a possible number of students.

**(b)** (i) The table gives $N(s) = 4$ sections for $s = 75$ and $s = 50$. For any integer between those in the table, the section number is the higher value. Therefore, for $50 \le s \le 75$, we have $N(s) = 4$ sections. We do not know what happens if $s < 50$.

(ii) First evaluate $N(125) = 5$. So we solve the equation $N(s) = 5$ for $s$. There are 5 sections when enrollment is between 76 and 125 students.

**37.** This represents the change in average hurricane intensity at current $CO_2$ levels if sea surface temperature rises by $1°C$.

## Solutions for Section 2.2

### Skill Refresher

**S1.** The function is undefined when the denominator is zero. Therefore, $x - 3 = 0$ tells us the function is undefined for $x = 3$.

**S5.** Adding 8 to both sides of the inequality we get $x > 8$.

**S9.** $x^2 - 25 > 0$ is true when $x > 5$ or $x < -5$.

### Exercises

**1.** The graph of $f(x) = 1/x$ for $-2 \leq x \leq 2$ is shown in Figure 2.1. From the graph, we see that $f(x) = -(1/2)$ at $x = -2$. As we approach zero from the left, $f(x)$ gets more and more negative. On the other side of the $y$-axis, $f(x) = (1/2)$ at $x = 2$. As $x$ approaches zero from the right, $f(x)$ grows larger and larger. Thus, on the domain $-2 \leq x \leq 2$, the range is $f(x) \leq -(1/2)$ or $f(x) \geq (1/2)$.

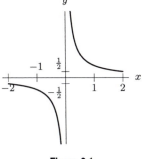

**Figure 2.1**

**5.** The domain is all real numbers except those which do not yield an output value. The expression $1/(x + 3)$ is defined for any real number $x$ except $-3$, since for $x = -3$ the denominator of $f(x)$, $x + 3$, is 0 and division by 0 is undefined. Therefore, the domain of $f(x)$ is all real numbers $\neq -3$.

**9.** To evaluate $f(x)$, we must have $x - 4 > 0$. Thus

$$\text{Domain: } x > 4.$$

To find the range, we want to know all possible output values. We solve the equation $y = f(x)$ for $x$ in terms of $y$. Since

$$y = \frac{1}{\sqrt{x - 4}},$$

squaring gives

$$y^2 = \frac{1}{x - 4},$$

and multiplying by $x - 4$ gives

$$y^2(x - 4) = 1$$
$$y^2 x - 4y^2 = 1$$
$$y^2 x = 1 + 4y^2$$
$$x = \frac{1 + 4y^2}{y^2}.$$

This formula tells us how to find the $x$-value which corresponds to a given $y$-value. The formula works for any $y$ except $y = 0$ (which puts a 0 in the denominator). We know that $y$ must be positive, since $\sqrt{x - 4}$ is positive, so we have

$$\text{Range: } y > 0.$$

**13.** Any number can be squared, so the domain is all real numbers.

**17.** $f(x)$ is not defined when $x = a$. So $a = 3$.

## Problems

**21.** The domain is $1 \leq x \leq 7$. The range is $2 \leq f(x) \leq 18$.

**25.** Since the restaurant opens at 2 pm, $t = 0$, and closes at 2 am, $t = 12$, a reasonable domain is $0 \leq t \leq 12$.

Since there cannot be fewer than 0 clients in the restaurant and 200 can fit inside, the range is $0 \leq f(t) \leq 200$.

**29. (a)** From the table we find that a 200 lb person uses 5.4 calories per minute while walking. So a half-hour, or a 30 minute, walk burns $30(5.4) = 162$ calories.

   **(b)** The number of calories used per minute is approximately proportional to the person's weight. The relationship is an approximately linear increasing function, where weight is the independent variable and number of calories burned is the dependent variable.

   **(c)** (i) Since the function is approximately linear, its equation is $c = b + mw$, where $c$ is the number of calories and $w$ is weight. Using the first two values in the table, the slope is

$$m = \frac{3.2 - 2.7}{120 - 100} = \frac{0.5}{20} = 0.025 \text{ cal/lb.}$$

Using the point $(100, 2.7)$ we have

$$2.7 = b + 0.025(100)$$
$$b = 0.2.$$

So the equation is $c = 0.2 + 0.025w$. See Figure 2.2. All the values given lie on this line with the exception of the last two which are slightly above it.

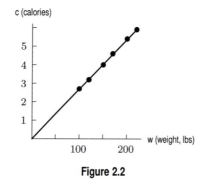

**Figure 2.2**

   (ii) The intercept $(0, 0.2)$ is the number of calories burned by a weightless runner. Since 0.2 is a small number, most of the calories burned appear to be due to moving a person's weight. Other methods of finding the equation of the the line may give other values for the vertical intercept, but all values are close to 0.

   (iii) Domain $0 < w$; range $0 < c$

   (iv) Evaluating the function at $w = 135$,

$$\text{Calories} = 0.2 + 0.025(135) \approx 3.6.$$

**33. (a)** Substituting $t = 0$ into the formula for $p(t)$ shows that $p(0) = 50$, meaning that there were 50 rabbits initially. Using a calculator, we see that $p(10) \approx 131$, which tells us there were about 131 rabbits after 10 months. Similarly, $p(50) \approx 911$ means there were about 911 rabbits after 50 months.

**(b)** The graph in Figure 2.3 tells us that the rabbit population grew quickly at first but then leveled off at about 1000 rabbits after around 75 months or so. It appears that the rabbit population increased until it reached the island's capacity.

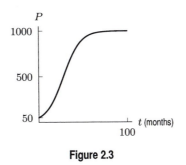

**Figure 2.3**

**(c)** From the graph in Figure 2.3, we see that the range is $50 \leq p(t) \leq 1000$. This tells us that (for $t \geq 0$) the number of rabbits is no less than 50 and no more than 1000.

**(d)** The smallest population occurred when $t = 0$. At that time, there were 50 rabbits. As $t$ gets larger and larger, $(0.9)^t$ gets closer and closer to 0. Thus, as $t$ increases, the denominator of

$$p(t) = \frac{1000}{1 + 19(0.9)^t}$$

decreases. As $t$ increases, the denominator $1 + 19(0.9)^t$ gets close to 1 (try $t = 100$, for example). As the denominator gets closer to 1, the fraction gets closer to 1000. Thus, as $t$ gets larger and larger, the population gets closer and closer to 1000. Thus, the range is $50 \leq p(t) < 1000$.

## Solutions for Section 2.3

### Skill Refresher

**S1.** Since the point zero is not included, this graph represents $x > 0$.

**S5.** Since both end points of the interval are solid dots, this graph represents $x \leq -1$ or $x \geq 2$.

**S9.** Domain: $-2 \leq x \leq 3$ and range: $-2 \leq x \leq 3$.

### Exercises

**1.** $f(x) = \begin{cases} -1, & -1 \leq x < 0 \\ 0, & 0 \leq x < 1 \\ 1, & 1 \leq x < 2 \end{cases}$    is shown in Figure 2.4.

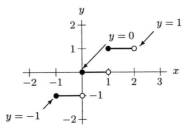

**Figure 2.4**

5.  Since $G(x)$ is defined for all $x$, the domain is all real numbers. For $x < -1$ the values of the function are all negative numbers. For $-1 \geq x \geq 0$ the functions values are $4 \geq G(x) \geq 3$, while for $x > 0$ we see that $G(x) \geq 3$ and the values increase to infinity. The range is $G(x) < 0$ and $G(x) \geq 3$.

9.  We find the formulas for each of the lines. For the first, we use the two points we have, $(1, 3.5)$ and $(3, 2.5)$. We find the slope: $(2.5 - 3.5)/(3 - 1) = -\frac{1}{2}$. Using the slope of $-\frac{1}{2}$, we solve for the $y$-intercept:

$$3.5 = b - \frac{1}{2} \cdot 1$$
$$4 = b.$$

Thus, the first line is $y = 4 - \frac{1}{2}x$, and it is for the part of the function where $1 \leq x \leq 3$.

We follow the same method for the second line, using the points $(5, 1)$ and $(8, 7)$. We find the slope: $(7-1)/(8-5) = 2$. Using the slope of 2, we solve for the $y$-intercept:

$$1 = b + 2 \cdot 5$$
$$-9 = b.$$

Thus, the second line is $y = -9 + 2x$, and it is for the part of the function where $5 \leq x \leq 8$.

Therefore, the function is:

$$y = \begin{cases} 4 - \frac{1}{2}x & \text{for } 1 \leq x \leq 3 \\ -9 + 2x & \text{for } 5 \leq x \leq 8. \end{cases}$$

## Problems

13. **(a)** Figure 2.5 shows the function $u(x)$. Some graphing calculators or computers may show a near vertical line close to the origin. The function seems to be $-1$ when $x < 0$ and 1 when $x > 0$.

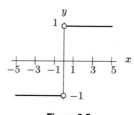

**Figure 2.5**

**(b)** Table 2.1 is the completed table. It agrees with what we found in part (a). The function is undefined at $x = 0$.

**Table 2.1**

| $x$ | $-5$ | $-4$ | $-3$ | $-2$ | $-1$ | 0 | 1 | 2 | 3 | 4 | 5 |
|---|---|---|---|---|---|---|---|---|---|---|---|
| $|x|/x$ | $-1$ | $-1$ | $-1$ | $-1$ | $-1$ | | 1 | 1 | 1 | 1 | 1 |

**(c)** The domain is all $x$ except $x = 0$. The range is $-1$ and $1$.

**(d)** $u(0)$ is undefined, not 0. The claim is false.

17. **(a)** The smaller the difference, the smaller the refund. The smallest possible difference is \$0.01. This translates into a refund of \$1.00 + \$0.01 = \$1.01.

**(b)** Looking at the refund rules, we see that there are three separate cases to consider. The first case is when 10 times the difference is less than \$1. If the difference is more than 0 but less than 10¢, and you will receive \$1 plus the difference. The formula for this is:

$$y = 1 + x \quad \text{for} \quad 0 < x < 0.10.$$

In the second case, 10 times the difference is between \$1 and \$5. This will be true if the difference is between 10¢ and 50¢. The formula for this is:

$$y = 10x + x \quad \text{for} \quad 0.10 \le x \le 0.50.$$

In the third case, 10 times the difference is more than \$5. If the difference is more than 50¢, then you receive \$5 plus the difference or:

$$y = 5 + x \quad \text{for} \quad x > 0.50.$$

Putting these cases together, we get:

$$y = \begin{cases} 1 + x & \text{for } 0 < x < 0.1 \\ 10x + x & \text{for } 0.1 \le x \le 0.5 \\ 5 + x & \text{for } x > 0.5. \end{cases}$$

(c) We want $x$ such that $y = 9$. Since the highest possible value of $y$ for the first case occurs when $x = 0.09$, and $y = 1 + 0.09 = \$1.09$, the range for this case does not go high enough. The highest possible value for the second case occurs when $x = 0.5$, and $y = 10(0.5) + 0.5 = \$5.50$. This range is also not high enough. So we look to the third case where $x > 0.5$ and $y = 5 + x$. Solving $5 + x = 9$ we find $x = 4$. So the price difference would have to be \$4.

(d) See Figure 2.6.

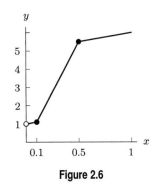

**Figure 2.6**

**21. (a)** We have $f(x) = \begin{cases} 2x - 6 & \text{for} \quad 2x - 6 \ge 0 \\ -(2x - 6) & \text{for} \quad 2x - 6 < 0 \end{cases}$

or

$$f(x) = \begin{cases} 2x - 6 & \text{for} \quad x \ge 3 \\ 6 - 2x & \text{for} \quad x < 3 \end{cases}.$$

**(b)** See Figure 2.7.

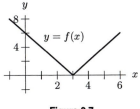

**Figure 2.7**

**25.** The statement

$$y = x^2 \text{ for } x \text{ less than zero, and } y = x - 1 \text{ for } x \text{ greater than or equal to zero}$$

can be condensed to $y = \begin{cases} x^2 & \text{for} \quad x < 0 \\ x - 1 & \text{for} \quad x \ge 0 \end{cases}$

# Solutions for Section 2.4

## Skill Refresher

**S1.** Adding 4 to both sides and dividing by 3 we get $y = \frac{x+4}{3}$.

**S5.** Add 4 to both sides and take the cube root to obtain $y = \sqrt[3]{x + 4}$.

**S9.** We have

$$3(y - 2)^2 - 7 = 3(y^2 - 4y + 4) - 7$$
$$= 3y^2 - 12y + 12 - 7$$
$$= 3y^2 - 12y + 5.$$

## Exercises

**1.** $A(f(t))$ is the area, in square centimeters, of the circle at time $t$ minutes.

**5.** $g(f(0)) = g(3 \cdot 0 - 1) = g(-1) = 1 - (-1)^2 = 0.$

**9.** $f(f(x)) = f(3x - 1) = 3(3x - 1) - 1 = 9x - 4.$

**13.** The inverse function, $f^{-1}(N)$, is the number of days for $N$ inches of snow to fall. Units of $f^{-1}(N)$ are days.

**17.** Since $Q = x^3 + 3$, solving for $x$ gives

$$x^3 + 3 = Q$$
$$x^3 = Q - 3$$
$$x = (Q - 3)^{1/3}$$
$$f^{-1}(Q) = (Q - 3)^{1/3}$$

**21. (a)** Since the vertical intercept of the graph of $f$ is $(0, b)$, we have $f(0) = b$.

**(b)** Since the horizontal intercept of the graph of $f$ is $(a, 0)$, we have $f(a) = 0$.

**(c)** The function $f^{-1}$ goes from $y$-values to $x$-values, so to evaluate $f^{-1}(0)$, we want the $x$-value corresponding to $y = 0$. This is $x = a$, so $f^{-1}(0) = a$.

**(d)** Solving $f^{-1}(?) = 0$ means finding the $y$-value corresponding to $x = 0$. This is $y = b$, so $f^{-1}(b) = 0$.

## Problems

**25. (a)** To find values of $f$, read the table from top to bottom, so

    (i)   $f(0) = 2$                         (ii)   $f(1) = 0.$

To find values of $f^{-1}$, read the table in the opposite direction (from bottom to top), so

    (iii)   $f^{-1}(0) = 1$                 (iv)   $f^{-1}(2) = 0.$

**(b)** Since the values $(0, 2)$ are paired in the table, we know $f(0) = 2$ and $f^{-1}(2) = 0$. Thus, knowing the answer to (i) (namely, $f(0) = 2$) tells us the answer to (iv). Similarly, the answer to (ii), namely $f(1) = 0$, tells us that the values $(1, 0)$ are paired in the table, so $f^{-1}(0) = 1$ too.

**29. (a)** $f(3) = 4 \cdot 3 = 12$ is the perimeter of a square of side 3.

**(b)** $f^{-1}(20)$ is the side of a square of perimeter 20. If $20 = 4s$, then $s = 5$, so $f^{-1}(20) = 5$.

**(c)** To find $f^{-1}(P)$, solve for $s$:

$$P = 4s$$
$$s = \frac{P}{4}$$
$$f^{-1}(P) = \frac{P}{4}.$$

**33.** We have

$$H = f(g(n)) = f(68 + 10 \cdot 2^{-n}) = \frac{5}{9}\left(68 + 10 \cdot 2^{-n} - 32\right) = 20 + \frac{50}{9}2^{-n},$$

and $f(g(n))$ gives the temperature, $H$, in degrees Celsius after $n$ hours.

**37.** Since $V = \frac{4}{3}\pi r^3$ and $r = 50 - 2.5t$, substituting $r$ into $V$ gives

$$V = f(t) = \frac{4}{3}\pi(50 - 2.5t)^3.$$

**41.** **(a)** $f(2) = 2.80$ means that 2 pounds of apples cost \$2.80.
   **(b)** $f(0.5) = 0.70$ means that 1/2 pound of apples cost \$0.70.
   **(c)** $f^{-1}(0.35) = 0.25$ means that \$0.35 buys 1/4 pound of apples.
   **(d)** $f^{-1}(7) = 5$ means that \$7 buys 5 pounds of apples.

**45.** Since $y = x^3 + 1$, solving for $x$ gives

$$x^3 + 1 = y$$
$$x^3 = y - 1$$
$$x = (y - 1)^{1/3}$$
$$f^{-1}(y) = (y - 1)^{1/3}.$$

**49.** To evaluate $p(x)$, we must have $3 - x > 0$. So the domain of $p(x)$ is all real numbers $< 3$. To find the range of $p(x)$, we find the inverse function of $p(x)$. Let $y = p(x)$. Solving for $x$, we get

$$y = \frac{1}{\sqrt{3 - x}}$$
$$\sqrt{3 - x} = \frac{1}{y}$$
$$3 - x = \frac{1}{y^2}$$
$$x = 3 - \frac{1}{y^2}$$
$$p^{-1}(y) = 3 - \frac{1}{y^2}.$$

The formula works for any $y$ except $y = 0$. We know that $y$ must be positive, since $\sqrt{3 - x}$ is positive, so the range of $p(x)$ is all real numbers $> 0$.

## Solutions for Section 2.5

## Exercises

**1.** To determine concavity, we calculate the rate of change:

$$\frac{\Delta f(x)}{\Delta x} = \frac{1.3 - 1.0}{1 - 0} = 0.3$$

$$\frac{\Delta f(x)}{\Delta x} = \frac{1.7 - 1.3}{3 - 1} = 0.2$$

$$\frac{\Delta f(x)}{\Delta x} = \frac{2.2 - 1.7}{6 - 3} \approx 0.167.$$

The rates of change are decreasing, so we expect the graph of $f(x)$ to be concave down.

**5.** The slope of $y = x^2$ is always increasing, so its graph is concave up. See Figure 2.8.

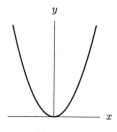

**Figure 2.8**

**9.** The rate of change between $t = 1.5$ and $t = 2.4$ is

$$\frac{\Delta R(t)}{\Delta t} = \frac{-3.1 - (-5.7)}{2.4 - 1.5} = 2.889.$$

Similarly, we have

$$\frac{\Delta R(t)}{\Delta t} = \frac{-1.4 - (-3.1)}{3.6 - 2.4} = 1.417$$

$$\frac{\Delta R(t)}{\Delta t} = \frac{0 - (-1.4)}{4.8 - 3.6} = 1.167.$$

The rate of change is decreasing, so we expect the graph to be concave down.

## Problems

**13.** This function is increasing throughout and the rate of increase is increasing, so the graph is concave up.

**17.** Since new people are always trying the product, it is an increasing function. At first, the graph is concave up. After many people start to use the product, the rate of increase slows down and the graph becomes concave down.

**21.** The graphical representation of the data is misleading because in the graph the number of violent crimes is put on the horizontal axis which give the graph the appearance of leveling out. This can fool us into believing that crime is leveling out. Note that it took from 1998 to 2000, about 2 years, for the number of violent crimes to go from 500 to 1,000, but it took less than 1/2 a year for that number to go from 1,500 to 2,000, and even less time for it to go from 2,000 to 2,500. In actuality, this graph shows that crime is growing at an increasing rate. If we were to graph the number of crimes as a function of the year, the graph would be concave up.

## Solutions for Chapter 2 Review

## Exercises

**1.** To evaluate when $x = -7$, we substitute $-7$ for $x$ in the function, giving $f(-7) = -\dfrac{7}{2} - 1 = -\dfrac{9}{2}$.

**5.** We have

$$y = f(4) = 4 \cdot 4^{3/2} = 4 \cdot 2^3 = 4 \cdot 8 = 32.$$

Solve for $x$:

$$4x^{3/2} = 6$$
$$x^{3/2} = 6/4$$
$$x^3 = 36/16 = 9/4$$
$$x = \sqrt[3]{9/4}.$$

**9.** Substituting 4 for $t$ gives

$$P(4) = 170 - 4 \cdot 4 = 154.$$

Similarly, with $t = 2$,

$$P(2) = 170 - 4 \cdot 2 = 162,$$

so

$$P(4) - P(2) = 154 - 162 = -8.$$

**13.** The expression $x^2 - 9$, found inside the square root sign, must always be non-negative. This happens when $x \geq 3$ or $x \leq -3$, so our domain is $x \geq 3$ or $x \leq -3$.
For the range, the smallest value $\sqrt{x^2 - 9}$ can have is zero. There is no largest value, so the range is $q(x) \geq 0$.

**17.** Since $m(t)$ is a linear function, the domain of $m(t)$ is all real numbers. For any value of $m(t)$ there is a corresponding value of $t$. So the range is also all real numbers.

**21.** (a) $2f(x) = 2(1 - x)$.
    (b) $f(x) + 1 = (1 - x) + 1 = 2 - x$.
    (c) $f(1 - x) = 1 - (1 - x) = x$.
    (d) $(f(x))^2 = (1 - x)^2$.
    (e) $f(1)/x = (1 - 1)/x = 0$.
    (f) $\sqrt{f(x)} = \sqrt{1 - x}$.

**25.** $P(f(t))$ is the period, in seconds, of the pendulum at time $t$ minutes.

**29.** $g(f(1)) = g(1^2 + 1) = g(2) = 2 \cdot 2 + 3 = 7$.

**33.** $g(g(x)) = g(2x + 3) = 2(2x + 3) + 3 = 4x + 9$.

**37.** Since $(x - b)^{1/2} = \sqrt{x - b}$, we know that $x - b \geq 0$. Thus, $x \geq b$. If $x = b$, then $(x - b)^{1/2} = 0$, which is the minimum value of $\sqrt{x - b}$, since it can't be negative. Thus, the range is all real numbers greater than or equal to 6.

**41.** If $P = f(t)$, then $t = f^{-1}(P)$, so $f^{-1}(P)$ gives the time in years at which the population is $P$ million.

**45.** We find $f(16) = 12 - \sqrt{16} = 8$. If $3 = 12 - \sqrt{x}$, then $\sqrt{x} = 9$, so $x = 81$. Thus, $f^{-1}(3) = 81$

## Problems

**49.** (a) Substituting, $q(5) = 3 - (5)^2 = -22$.
    (b) Substituting, $q(a) = 3 - a^2$.
    (c) Substituting, $q(a - 5) = 3 - (a - 5)^2 = 3 - (a^2 - 10a + 25) = -a^2 + 10a - 22$.
    (d) Using the answer to part (b), $q(a) - 5 = 3 - a^2 - 5 = -a^2 - 2$.
    (e) Using the answer to part (b) and (a), $q(a) - q(5) = (3 - a^2) - (-22) = -a^2 + 25$.

**53.** (a) Substituting $x = 0$ gives $f(0) = \sqrt{0^2 + 16} - 5 = \sqrt{16} - 5 = 4 - 5 = -1$.
    (b) We want to find $x$ such that $f(x) = \sqrt{x^2 + 16} - 5 = 0$. Thus, we have

$$\sqrt{x^2 + 16} - 5 = 0$$
$$\sqrt{x^2 + 16} = 5$$
$$x^2 + 16 = 25$$
$$x^2 = 9$$
$$x = \pm 3.$$

    Thus, $f(x) = 0$ for $x = 3$ or $x = -3$.
    (c) In part (b), we saw that $f(3) = 0$. You can verify this by substituting $x = 3$ into the formula for $f(x)$:

$$f(3) = \sqrt{3^2 + 16} - 5 = \sqrt{25} - 5 = 5 - 5 = 0.$$

    (d) The vertical intercept is the value of the function when $x = 0$. We found this to be $-1$ in part (a). Thus the vertical intercept is $-1$.
    (e) The graph touches the $x$-axis when $f(x) = 0$. We saw in part (b) that this occurs at $x = 3$ and $x = -3$.

**57. (a)** To write $s$ as a function of $A$, we solve $A = 6s^2$ for $s$

$$s^2 = \frac{A}{6} \qquad \text{so} \qquad s = f(A) = +\sqrt{\frac{A}{6}} \qquad \text{Because the length of a side of a cube is positive.}$$

The function $f$ gives the side of a cube in terms of its area $A$.

**(b)** Substituting $s = f(A) = \sqrt{A/6}$ in the formula $V = g(s) = s^3$ gives the volume, $V$, as a function of surface area, $A$,

$$V = g(f(A)) = s^3 = \left(\sqrt{\frac{A}{6}}\right)^3.$$

**61. (a)** Using Pythagoras' Theorem, we see that the diagonal $d$ is given in terms of $s$ by

$$d^2 = 2s^2$$
$$s = \sqrt{\frac{d^2}{2}} = \frac{d}{\sqrt{2}}$$
$$s = f(d) = \frac{d}{\sqrt{2}}.$$

**(b)** $A = g(s) = s^2$.

**(c)** Substituting $s = d/\sqrt{2}$ in $g$ gives

$$A = g(s) = \left(\frac{d}{\sqrt{2}}\right)^2 = \frac{d^2}{2}.$$

**(d)** The function $h$ is the composition of $f$ and $g$, with $f$ as the inside function, that is $h(d) = g(f(d))$.

**65. (a)** $t(400) = 272$.

**(b)**   (i) It takes 136 seconds to melt 1 gram of the compound at a temperature of $800°C$.

   (ii) It takes 68 seconds to melt 1 gram of the compound at a temperature of $1600°C$.

**(c)** This means that $t(2x) = t(x)/2$, because if $x$ is a temperature and $t(x)$ is a melting time, then $2x$ would be double this temperature and $t(x)/2$ would be half this melting time.

**69.**   • A function such as $y = \sqrt{x - 4}$ is undefined for $x < 4$, because the input of the square root operation is negative for these $x$-values.

   • A function such as $y = 1/(x - 8)$ is undefined for $x = 8$.

   • Combining two functions such as these, for example by adding or multiplying them, yields a function with the required domain. Thus, possible formulas include

$$y = \frac{1}{x - 8} + \sqrt{x - 4} \qquad \text{or} \qquad y = \frac{\sqrt{x - 4}}{x - 8}.$$

## CHECK YOUR UNDERSTANDING

**1.** False. $f(2) = 3 \cdot 2^2 - 4 = 8$.

**5.** False. $W = (8 + 4)/(8 - 4) = 3$.

**9.** True. A fraction can only be zero if the numerator is zero.

**13.** False. The domain consists of all real numbers $x$, $x \neq 3$

**17.** True. Since $f$ is an increasing function, the domain endpoints determine the range endpoints. We have $f(15) = 12$ and $f(20) = 14$.

**21.** True. $|x| = |-x|$ for all $x$.

**25.** True. If $x < 0$, then $f(x) = x < 0$, so $f(x) \neq 4$. If $x > 4$, then $f(x) = -x < 0$, so $f(x) \neq 4$. If $0 \leq x \leq 4$, then $f(x) = x^2 = 4$ only for $x = 2$. The only solution for the equation $f(x) = 4$ is $x = 2$.

**29.** True. To find $f^{-1}(R)$, we solve $R = \frac{2}{3}S + 8$ for $S$ by subtracting 8 from both sides and then multiplying both sides by $(3/2)$.

**33.** False. Since

$$f(g(x)) = 2\left(\frac{1}{2}x - 1\right) + 1 = x - 1 \neq x,$$

the functions do not undo each other.

**37.** True. Since the function is concave up, the average rate of change increases as we move right.

**41.** True. For $x > 0$, the function $f(x) = -x^2$ is both decreasing and concave down.

# CHAPTER THREE

## Solutions for Section 3.1

### Skill Refresher

**S1.** In this example, we distribute the factors $50t$ and $2t$ across the two binomials $t^2 + 1$ and $25t^2 + 125$, respectively. Thus,

$$(t^2 + 1)(50t) - (25t^2 + 125)(2t) = 50t^3 + 50t - (50t^3 + 250t)$$
$$= 50t^3 + 50t - 50t^3 - 250t = -200t.$$

**S5.** $3x^2 - x - 4 = (3x - 4)(x + 1)$

**S9.**
$$x^2 + 7x + 6 = 0$$
$$(x + 6)(x + 1) = 0$$
$$x + 6 = 0 \quad \text{or} \quad x + 1 = 0$$
$$x = -6 \quad \text{or} \quad x = -1$$

### Exercises

**1.** Yes. We rewrite the function giving

$$f(x) = 2(7 - x)^2 + 1$$
$$= 2(49 - 14x + x^2) + 1$$
$$= 98 - 28x + 2x^2 + 1$$
$$= 2x^2 - 28x + 99.$$

So $f(x)$ is quadratic with $a = 2$, $b = -28$ and $c = 99$.

**5.** No. We rewrite the function giving

$$R(q) = \frac{1}{q^2}(q^2 + 1)^2$$
$$= \frac{1}{q^2}(q^4 + 2q^2 + 1)$$
$$= q^2 + 2 + \frac{1}{q^2}$$
$$= q^2 + 2 + q^{-2}.$$

So $R(q)$ is not quadratic since it contains a term with $q$ to a negative power.

**9.** We solve for $x$ in the equation $5x - x^2 + 3 = 0$ using the quadratic formula with $a = -1$, $b = 5$ and $c = 3$.

$$x = \frac{-5 \pm \sqrt{(-5)^2 - 4(-1)3}}{2(-1)}$$
$$x = \frac{-5 \pm \sqrt{37}}{-2}.$$

The solutions are $x \approx -0.541$ and $x \approx 5.541$.

**13.** To find the zeros, we solve the equation

$$0 = 4x^2 - 4x - 8.$$

We see that this is factorable, as follows:

$$0 = 4(x^2 - x - 2)$$
$$0 = 4(x - 2)(x + 1).$$

Therefore, the zeros occur where $x = 2$ and $x = -1$.

**17.** Letting $z = x^2$, we have $y = z^2 + 5z + 6$. This can be factored, giving

$$y = (z + 2)(z + 3)$$
$$= (x^2 + 2)(x^2 + 3).$$

Setting the factors equal to zero, we have $x^2 + 2 = 0$, which has no solution, and $x^2 + 3 = 0$, which also has no solution, so this function has no real-valued zeros. Another way to see this is to notice that both $x^4$ and $5x^2$ are either positive or 0, so $y$ can not be less than 6.

## Problems

**21.** We solve the equation $f(t) = -16t^2 + 47t + 3 = 0$ using the quadratic formula

$$-16t^2 + 47t + 3 = 0$$
$$t = \frac{-47 \pm \sqrt{47^2 - 4(-16)3}}{2(-16)}.$$

Evaluating gives $t = -1/16$ sec and $t = 3$ sec; the value $t = 3$ sec is the time we want. The baseball hits the ground 3 sec after it was hit.

**25.** For example, we can use $y = (x + 2)(x - 3)$. See Figure 3.1.

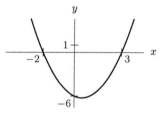

**Figure 3.1**

**29.** We know there are zeros at $x = -6$ and $x = 2$, so we use the factored form:

$$y = a(x + 6)(x - 2)$$

and solve for $a$. At $x = 0$, we have

$$5 = a(0 + 6)(0 - 2)$$
$$5 = -12a$$
$$-\frac{5}{12} = a.$$

Thus,

$$y = -\frac{5}{12}(x + 6)(x - 2)$$

or

$$y = -\frac{5}{12}x^2 - \frac{5}{3}x + 5.$$

**33.** **(a)** The initial velocity is the velocity when $t = 0$. So $v(0) = 0^2 - 4 \cdot 0 + 4 = 4$ meters per second.
  **(b)** The object is not moving when its velocity is zero. This time is found by factoring $t^2 - 4t + 4 = (t-2)^2 = 0$. The solution, $t = 2$, tells us that the object is not moving at 2 seconds.
  **(c)** From the graph of the velocity function in Figure 3.2, we can see that it is concave up.

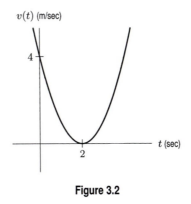

**Figure 3.2**

**37.** **(a)** According to the figure in the text, the package was dropped from a height of 5 km.
  **(b)** When the package hits the ground, $h = 0$ and $d = 4430$. So, the package has moved 4430 meters forward when it lands.
  **(c)** Since the maximum is at $d = 0$, the formula is of the form $h = ad^2 + b$ where a is negative and b is positive. Since $h = 5$ at $d = 0$, $5 = a(0)^2 + b = b$, so $b = 5$. We now know that $h = ad^2 + 5$. Since $h = 0$ when $d = 4430$, we have $0 = a(4430)^2 + 5$, giving $a = \dfrac{-5}{(4430)^2} \approx -0.000000255$. So $h \approx -0.000000255d^2 + 5$.

# Solutions for Section 3.2

## Skill Refresher

**S1.** $y^2 - 12y = y^2 - 12y + 36 - 36 = (y-6)^2 - 36$

**S5.** Get the variables on the left side, the constants on the right side and complete the square using $\left(\frac{-6}{2}\right)^2 = 9$.

$$r^2 - 6r = -8$$
$$r^2 - 6r + 9 = 9 - 8$$
$$(r-3)^2 = 1.$$

Take the square root of both sides and solve for $r$.

$$r - 3 = \pm 1$$
$$r = 3 \pm 1.$$

So, $r = 4$ or $r = 2$.

**S9.** Rewrite the equation to equal zero, and factor.

$$n^2 + 4n - 5 = 0$$
$$(n+5)(n-1) = 0.$$

So, $n + 5 = 0$ or $n - 1 = 0$, thus $n = -5$ or $n = 1$.

## Exercises

**1.** By comparing $f(x)$ to the vertex form, $y = a(x - h)^2 + k$, we see the vertex is $(h, k) = (1, 2)$. The axis of symmetry is the vertical line through the vertex, so the equation is $x = 1$. The parabola opens upward because the value of $a$ is positive
3.

**5. (a)** See Figure 3.3. For $g$, we have $a = 1$, $b = 0$, and $c = 3$. Its vertex is at $(0, 3)$, and its axis of symmetry is the $y$-axis, or the line $x = 0$. This function has no zeros.

**(b)** See Figure 3.4. For $f$, we have $a = -2$, $b = 4$, and $c = 16$. The axis of symmetry is the line $x = 1$ and the vertex is at $(1, 18)$. The zeros, or $x$-intercepts, are at $x = -2$ and $x = 4$. The $y$-intercept is at $y = 16$.

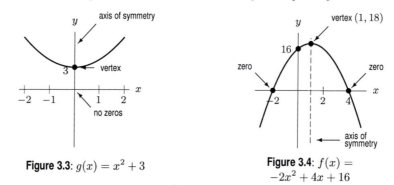

**Figure 3.3**: $g(x) = x^2 + 3$    **Figure 3.4**: $f(x) = -2x^2 + 4x + 16$

**9.** Since the vertex is $(4, 7)$, we use the form $y = a(x - h)^2 + k$, with $h = 4$ and $k = 7$. We solve for $a$, substituting in the second point, $(0, 4)$.

$$y = a(x - 4)^2 + 7$$
$$4 = a(0 - 4)^2 + 7$$
$$-3 = 16a$$
$$-\frac{3}{16} = a.$$

Thus, an equation for the parabola is

$$y = -\frac{3}{16}(x - 4)^2 + 7.$$

**13.** The square of half the coefficient of the $x$-term is $\left(\frac{8}{2}\right)^2 = 16$. Adding and subtracting this number after the $x$-term gives

$$f(x) = x^2 + 8x + 16 - 16 + 3.$$

This can be simplified to $f(x) = (x + 4)^2 - 13$. The vertex is $(-4, -13)$ and the axis of symmetry is $x = -4$.

**17.** The standard form is obtained by writing the right-hand side as three terms and rearranging the terms:

$$y = \frac{1}{2}x^2 - \frac{1}{2}x - 6.$$

The vertex form can be obtained from the standard form by completing the square:

$$y = \frac{1}{2}x^2 - \frac{1}{2}x - 6$$
$$= \frac{1}{2}(x^2 - x - 12)$$
$$= \frac{1}{2}(x^2 - x + \frac{1}{4} - \frac{1}{4} - 12)$$
$$= \frac{1}{2}(x^2 - x + \frac{1}{4}) + \frac{1}{2}(-\frac{1}{4} - 12)$$

$$= \frac{1}{2}\left(x - \frac{1}{2}\right)^2 + \frac{1}{2}\left(-\frac{49}{4}\right)$$
$$= \frac{1}{2}\left(x - \frac{1}{2}\right)^2 - \frac{49}{8}.$$

The factored form can be obtained by factoring:

$$y = \frac{1}{2}(x^2 - x - 12)$$
$$= \frac{1}{2}(x - 4)(x + 3).$$

## Problems

**21.** We have $(h, k) = (4, 2)$, so $y = a(x - 4)^2 + 2$. Solving for $a$, we have

$$a(0 - 4)^2 + 2 = 6$$
$$16a = 4$$
$$a = \frac{1}{4},$$

so $y = (1/4)(x - 4)^2 + 2$.

**25. (a)**

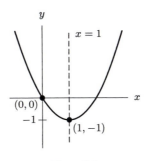

**Figure 3.5**

**(b)** Since the vertex is at $(1, -1)$, the parabola could be described by

$$f(x) = a(x - 1)^2 - 1.$$

Since the parabola passes through $(0, 0)$

$$0 = a(0 - 1)^2 - 1$$
$$0 = a - 1.$$

Therefore

$$a = 1.$$

So, the equation is

$$f(x) = (x - 1)^2 - 1$$

or

$$f(x) = x^2 - 2x.$$

**(c)** Since the vertex is at $(1, -1)$ and the parabola is concave up, the range of this function is all real numbers greater than or equal to $-1$.

**(d)** Since one zero is at $x = 0$, which is one unit to the left of the axis of symmetry at $x = 1$, the other zero will occur at one unit to the right of the axis of the symmetry at $x = 2$.

**29.** The distance around any rectangle with a height of $h$ units and a base of $b$ units is $2b + 2h$. See Figure 3.6. Since the string forming the rectangle is 50 cm long, we know that $2b + 2h = 50$ or $b + h = 25$. Therefore, $b = 25 - h$. The area, $A$, of such a rectangle is

$$A = bh$$
$$A = (25 - h)(h).$$

The zeros of this quadratic function are $h = 0$ and $h = 25$, so the axis of symmetry, which is halfway between the zeros, is $h = 12.5$. Since the maximum value of $A$ occurs on the axis of symmetry, the area will be the greatest when the height is 12.5 and the base is also 12.5 ($b = 25 - h = 25 - 12.5 = 12.5$).

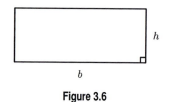

**Figure 3.6**

If the string were $k$ cm long, $2b + 2h = k$ or $b + h = \frac{k}{2}$, so $b = \frac{k}{2} - h$. $A = bh = \left(\frac{k}{2} - h\right)h$. The zeros in this case are $h = 0$ and $h = \frac{k}{2}$, so the axis of symmetry is $h = \frac{k}{4}$. If $h = \frac{k}{4}$, then $b = \frac{k}{2} - h = \frac{k}{2} - \frac{k}{4} = \frac{k}{4}$. So the dimensions for maximum area are $\frac{k}{4}$ by $\frac{k}{4}$; in other words, the rectangle with the maximum area is a square whose side measures $\frac{1}{4}$ of the length of the string.

# Solutions for Chapter 3 Review

## Exercises

**1.** We have:
$$f(x) = (2x - 3)(5 - x) = -2x^2 + 13x - 15,$$
so $a = -2, b = 13, c = -15$.

**5.** To find the zeros, we solve the equation
$$0 = 9x^2 + 6x + 1.$$

We see that this is factorable, as follows:
$$y = (3x + 1)(3x + 1)$$
$$y = (3x + 1)^2.$$

Therefore, there is only one zero at $x = -\frac{1}{3}$.

**9.** We solve for $t$ in the equation by factoring
$$N(t) = t^2 - 7t + 10 = (t - 2)(t - 5) = 0.$$

The zeros of N(t) are $t = 2$ and $t = 5$.

**13.** We have $y = a(x - 1)^2 - 2$ and if $x = 0$, $y = -5$, so $-5 = a(-1)^2 - 2$. Therefore, $a = -3$, and we have $y = -3(x - 1)^2 - 2$.

**17.** We have $y = a(x+1)(x-2)$. Since $(-2, 16)$ is on the curve, $16 = a(-1)(-4)$. Therefore $a = 4$, so $y = 4(x+1)(x-2)$.

**21.** The function appears quadratic with vertex at $(2, 0)$, so it could be of the form $y = a(x - 2)^2$. For $x = 0$, $y = -4$, so $y = a(0 - 2)^2 = 4a = -4$ and $a = -1$. Thus $y = -(x - 2)^2$ is a possible formula.

## Problems

**25.** By inspection the vertex is $(0.6, 0)$. The axis of symmetry is $x = 0.6$. The $y$-intercept occurs when $x = 0$, so $y = 0.36$. Since the coefficient $a$ is positive the curve is concave up.

**29.** We have $y = -3x^2 + 24x - 36 = -3(x^2 - 8x + 12) = -3(x - 6)(x - 2)$. The zeros are at $x = 6$ and $x = 2$. The axis of symmetry occurs midway between the zeros, so it is $x = 4$. The vertex is on the axis of symmetry and can be found by setting $x = 4$ and solving for $y$. This means that $y = -3(-2)(2) = 12$. Therefore the vertex is $(4, 12)$.

**33.** Figure 3.7 shows a graph of the basketball player's trajectory for $T = 1$ second. Since this is the graph of a parabola, the maximum height occurs at the $t$-value which is halfway between the zeros, 0 and 1. Thus, the maximum occurs at $t = 1/2$ second. The maximum height is $h(1/2) = 4$ feet, and 75% of 4 is 3. Thus, when the basketball player is above 3 feet from the ground, he is in the top 25% of his trajectory. To find when he reaches a height of 3 feet, set $h(t) = 3$. Solving for $t$ gives $t = 0.25$ or $t = 0.75$ seconds. Thus, from $t = 0.25$ to $t = 0.75$ seconds, the basketball player is in the top 25% of his jump, as indicated in Figure 3.7. We see that he spends half of the time at the top quarter of the height of this jump, giving the impression that he hangs in the air.

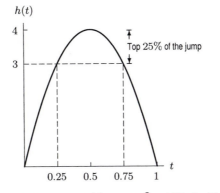

**Figure 3.7**: A graph of $h(t) = -16t^2 + 16Tt$ for $T = 1$

## CHECK YOUR UNDERSTANDING

**1.** True. It is of the form $f(x) = a(x - r)(x - s)$ where $a = 1$, $r = 0$ and $s = -2$.

**5.** False. The time when the object hits the ground is when the height is zero, $s(t) = 0$. The value $s(0)$ gives the height when the object is launched at $t = 0$.

**9.** False. A quadratic function may have two, one, or no zeros.

**13.** False. The vertex is located at the point $(h, k)$.

## Solutions to Skills for Factoring

**1.** $2(3x - 7) = 6x - 14$

**5.** $x(2x + 5) = 2x^2 + 5x$

**9.** $5z(x - 2) - 3(x - 2) = 5xz - 10z - 3x + 6$

**13.** $(y + 1)(z + 3) = yz + 3y + z + 3$

**17.** $(x-5)6 - 5(1-(2-x)) = 6x - 30 - 5(1-2+x) = 6x - 30 + 5 - 5x = x - 25.$

**21.** First we square $\sqrt{2x}+1$ and then take the negative of this result. Therefore,

$$-\left(\sqrt{2x}+1\right)^2 = -\left(\sqrt{2x}+1\right)\left(\sqrt{2x}+1\right) = -\left(2x + \sqrt{2x} + \sqrt{2x} + 1\right)$$
$$= -(2x + 2\sqrt{2x} + 1) = -2x - 2\sqrt{2x} - 1.$$

**25.** $5z - 30 = 5(z-6)$

**29.** $3u^7 + 12u^2 = 3u^2(u^5 + 4)$

**33.** $x^2 - 3x + 2 = (x-2)(x-1)$

**37.** Can be factored no further.

**41.** $x^2 + 3x - 28 = (x+7)(x-4)$

**45.** $x^2 + 2xy + 3xz + 6yz = x(x+2y) + 3z(x+2y) = (x+2y)(x+3z).$

**49.** We notice that the only factors of 24 whose sum is $-10$ are $-6$ and $-4$. Therefore,

$$B^2 - 10B + 24 = (B-6)(B-4).$$

**53.** This example is factored as the difference of perfect squares. Thus,

$$(t+3)^2 - 16 = ((t+3)-4)((t+3)+4)$$
$$= (t-1)(t+7).$$

Alternatively, we could arrive at the same answer by multiplying the expression out and then factoring it.

**57.**

$$c^2d^2 - 25c^2 - 9d^2 + 225 = c^2(d^2-25) - 9(d^2-25)$$
$$= (d^2-25)(c^2-9)$$
$$= (d+5)(d-5)(c+3)(c-3).$$

**61.** The common factor is $xe^{-3x}$. Therefore,

$$x^2 e^{-3x} + 2xe^{-3x} = xe^{-3x}(x+2).$$

**65.** $dk + 2dm - 3ek - 6em = d(k+2m) - 3e(k+2m) = (k+2m)(d-3e).$

**69.**

$$x = \frac{-3 \pm \sqrt{3^2 - 4(4)(-15)}}{2(4)}$$
$$x = \frac{-3 \pm \sqrt{249}}{8}$$

**73.**

$$-16t^2 + 96t + 12 = 60$$
$$-16t^2 + 96t - 48 = 0$$
$$t^2 - 6t + 3 = 0$$

$$t = \frac{-(-6) \pm \sqrt{(-6)^2 - 4(1)(3)}}{2(1)}$$
$$t = \frac{6 \pm \sqrt{24}}{2} = \frac{6 \pm 2\sqrt{6}}{2}$$
$$t = 3 \pm \sqrt{6}.$$

**77.**

$$N^2 - 2N - 3 = 2N(N - 3)$$
$$N^2 - 2N - 3 = 2N^2 - 6N$$
$$N^2 - 4N + 3 = 0$$
$$(N - 3)(N - 1) = 0$$
$$N = 3 \ \text{ or } \ N = 1$$

**81.** We rewrite the quadratic equation in standard form and use the quadratic formula. So

$$60 = -16t^2 + 96t + 12$$
$$16t^2 - 96t + 48 = 0$$
$$t^2 - 6t + 3 = 0$$
$$t = \frac{-(-6) \pm \sqrt{(-6)^2 - 4(1)(3)}}{2} = \frac{6 \pm \sqrt{36 - 12}}{2}$$
$$= \frac{6 \pm \sqrt{24}}{2} = \frac{6 \pm 2\sqrt{6}}{2} = 3 \pm \sqrt{6}.$$

**85.** To find the common denominator, we factor the second denominator

$$\frac{2}{z - 3} + \frac{7}{z^2 - 3z} = 0$$
$$\frac{2}{z - 3} + \frac{7}{z(z - 3)} = 0$$

which produces a common denominator of $z(z - 3)$. Therefore:

$$\frac{2z}{z(z - 3)} + \frac{7}{z(z - 3)} = 0$$
$$\frac{2z + 7}{z(z - 3)} = 0$$
$$2z + 7 = 0$$
$$z = -\frac{7}{2}.$$

**89.** We can solve this equation by squaring both sides.

$$\sqrt{r^2 + 24} = 7$$
$$r^2 + 24 = 49$$
$$r^2 = 25$$
$$r = \pm 5$$

**93.** Multiply by $(x - 5)(x - 1)$ on both sides of the equation, giving

$$(3x + 4)(x - 2) = 0.$$

So, $3x + 4 = 0$, or $x - 2 = 0$, that is,

$$x = -\frac{4}{3}, \quad x = 2.$$

**97.** For a fraction to equal zero, the numerator must equal zero. So, we solve

$$x^2 - 5mx + 4m^2 = 0.$$

Since $x^2 - 5mx + 4m^2 = (x - m)(x - 4m)$, we know that the numerator equals zero when $x = 4m$ and when $x = m$. But for $x = m$, the denominator will equal zero as well. So, the fraction is undefined at $x = m$, and the only solution is $x = 4m$.

**101.** We substitute the expression $4 - x^2$ for $y$ in the second equation.

$$y - 2x = 1$$
$$4 - x^2 - 2x = 1$$
$$-x^2 - 2x + 3 = 0$$
$$x^2 + 2x - 3 = 0$$
$$(x + 3)(x - 1) = 0$$
$$x = -3 \quad \text{and} \quad y = 4 - (-3)^2 = -5 \quad \text{or}$$
$$x = 1 \quad \text{and} \quad y = 4 - 1^2 = 3$$

**105.** Solving $y = x^2$ and $y = 15 - 2x$ simultaneously, we have

$$x^2 = 15 - 2x$$
$$x^2 + 2x - 15 = 0$$
$$(x + 5)(x - 3) = 0$$
$$x = -5, 3.$$

Thus, the points of intersection are $(-5, 25)$, $(3, 9)$.

## Solutions to Skills for Completing the Square

**1.** $x^2 + 8x = x^2 + 8x + 16 - 16 = (x + 4)^2 - 16$

**5.** We add and subtract the square of half the coefficient of the $a$-term, $\left(\frac{-2}{2}\right)^2 = 1$, to get

$$a^2 - 2a - 4 = a^2 - 2a + 1 - 1 - 4$$
$$= (a^2 - 2a + 1) - 1 - 4$$
$$= (a - 1)^2 - 5.$$

**9.** Completing the square yields

$$x^2 - 2x - 3 = (x^2 - 2x + 1) - 1 - 3 = (x - 1)^2 - 4.$$

**13.** Complete the square and write in vertex form.

$$y = x^2 + 6x + 3$$
$$= x^2 + 6x + 9 - 9 + 3$$
$$= (x + 3)^2 - 6.$$

The vertex is $(-3, -6)$.

**17.** Complete the square and write in vertex form.

$$
\begin{aligned}
y &= -x^2 + x - 6 \\
&= -(x^2 - x + 6) \\
&= -\left(x^2 - x + \frac{1}{4} - \frac{1}{4} + 6\right) \\
&= -\left(\left(x - \frac{1}{2}\right)^2 - \frac{1}{4} + 6\right) \\
&= -\left(x - \frac{1}{2}\right)^2 - \frac{23}{4}.
\end{aligned}
$$

The vertex is $(1/2, -23/4)$.

**21.** Complete the square and write in vertex form.

$$
\begin{aligned}
y &= 2x^2 - 7x + 3 \\
&= 2\left(x^2 - \frac{7}{2}x + \frac{3}{2}\right) \\
&= 2\left(x^2 - \frac{7}{2}x + \frac{49}{16} - \frac{49}{16} + \frac{3}{2}\right) \\
&= 2\left(\left(x - \frac{7}{4}\right)^2 - \frac{49}{16} + \frac{3}{2}\right) \\
&= 2\left(x - \frac{7}{4}\right)^2 - \frac{25}{8}.
\end{aligned}
$$

The vertex is $(7/4, -25/8)$.

**25.** Complete the square with $\left(\frac{1}{2}\right)^2 = \frac{1}{4}$ and take the square root of both sides to solve for $d$.

$$
\begin{aligned}
d^2 - d + \frac{1}{4} &= \frac{1}{4} + 2 \\
\left(d - \frac{1}{2}\right)^2 &= \frac{9}{4} \\
d - \frac{1}{2} &= \pm\frac{3}{2} \\
d &= \frac{1}{2} \pm \frac{3}{2}.
\end{aligned}
$$

So $d = 2$ or $d = -1$.

**29.** Complete the square on the left side.

$$
\begin{aligned}
5\left(p^2 + \frac{9}{5}p\right) &= 1 \\
5\left(p^2 + \frac{9}{5}p + \frac{81}{100}\right) &= 5\left(\frac{81}{100}\right) + 1 \\
5\left(p + \frac{9}{10}\right)^2 &= \frac{81}{20} + 1 \\
5\left(p + \frac{9}{10}\right)^2 &= \frac{101}{20}.
\end{aligned}
$$

Divide by 5 and take the square root of both sides to solve for $p$.

$$
\left(p + \frac{9}{10}\right)^2 = \frac{101}{100}
$$

$$p + \frac{9}{10} = \pm\sqrt{\frac{101}{100}}$$

$$p + \frac{9}{10} = \pm\frac{\sqrt{101}}{10}$$

$$p = -\frac{9}{10} \pm \frac{\sqrt{101}}{10}.$$

**33.** Set the equation equal to zero, $w^2 + w - 4 = 0$. With $a = 1$, $b = 1$, and $c = -4$, we use the quadratic formula,

$$w = \frac{-b \pm \sqrt{b^2 - 4ac}}{2a}$$

$$= \frac{-1 \pm \sqrt{1^2 - 4 \cdot 1 \cdot (-4)}}{2 \cdot 1}$$

$$= \frac{-1 \pm \sqrt{1 + 16}}{2}$$

$$= \frac{-1 \pm \sqrt{17}}{2}.$$

**37.** Solve by completing the square using $\left(\frac{3}{2}\right)^2 = \frac{9}{4}$.

$$s^2 + 3s + \frac{9}{4} = 1 + \frac{9}{4}$$

$$\left(s + \frac{3}{2}\right)^2 = 1 + \frac{9}{4}$$

$$\left(s + \frac{3}{2}\right)^2 = \frac{13}{4}.$$

Taking the square root of both sides and solving for $s$,

$$s + \frac{3}{2} = \pm\sqrt{\frac{13}{4}}$$

$$s + \frac{3}{2} = \pm\frac{\sqrt{13}}{2}$$

$$s = -\frac{3}{2} \pm \frac{\sqrt{13}}{2}$$

$$s = \frac{-3 \pm \sqrt{13}}{2}.$$

**41.** Simplify by dividing by 3 and solve by completing the square.

$$y^2 = 2y + 6$$

$$y^2 - 2y = 6$$

$$y^2 - 2y + 1 = 1 + 6$$

$$(y - 1)^2 = 7.$$

Take the square root of both sides and solve for $y$ to get $y = 1 \pm \sqrt{7}$.

**45.** Use the quadratic formula with $a = 49$, $b = 70$, $c = 22$, to solve this equation.

$$m = \frac{-70 \pm \sqrt{70^2 - 4 \cdot 49 \cdot 22}}{2 \cdot 49}$$

$$= \frac{-70 \pm \sqrt{4900 - 4312}}{98}$$

$$= \frac{-70 \pm \sqrt{588}}{98}$$

$$= \frac{-70 \pm 14\sqrt{3}}{98}$$

$$= \frac{-5 \pm \sqrt{3}}{7}.$$

# CHAPTER FOUR

## Solutions for Section 4.1

### Skill Refresher

**S1.** We have $6\% = 0.06$.

### Exercises

**1.** Yes. Writing the function as

$$g(w) = 2\left(2^{-w}\right) = 2\left(2^{-1}\right)^w = 2\left(\frac{1}{2}\right)^w,$$

we have $a = 2$ and $b = 1/2$.

**5.** Yes. Writing the function as

$$q(r) = \frac{-4}{3^r} = -4\left(\frac{1}{3^r}\right) = -4\left(\frac{1^r}{3^r}\right) = -4\left(\frac{1}{3}\right)^r,$$

we have $a = -4$ and $b = 1/3$.

**9.** No. The two terms cannot be combined into the form $b^r$.

**13.** The growth factor per century is $1+$ the growth per century. Since the forest is shrinking, the growth is negative, so we subtract 0.80, giving 0.20.

**17.** We can rewrite this as

$$Q = 0.0022(2.31^{-3})^t$$
$$= 0.0022(0.0811)^t,$$

so $a = 0.0022$, $b = 0.0811$, and $r = b - 1 = -0.9189 = -91.89\%$.

### Problems

**21.** To match formula and graph, we keep in mind the effect on the graph of the parameters $a$ and $b$ in $y = ab^t$.

If $a > 0$ and $b > 1$, then the function is positive and increasing.
If $a > 0$ and $0 < b < 1$, then the function is positive and decreasing.
If $a < 0$ and $b > 1$, then the function is negative and decreasing.
If $a < 0$ and $0 < b < 1$, then the function is negative and increasing.

**(a)** $y = 0.8^t$. So $a = 1$ and $b = 0.8$. Since $a > 0$ and $0 < b < 1$, we want a graph that is positive and decreasing. The graph in (ii) satisfies the conditions.
**(b)** $y = 5(3)^t$. So $a = 5$ and $b = 3$. The graph in (i) is both positive and increasing.
**(c)** $y = -6(1.03)^t$. So $a = -6$ and $b = 1.03$. Here, $a < 0$ and $b > 1$, so we need a graph which is negative and decreasing. The graph in (iv) satisfies these conditions.
**(d)** $y = 15(3)^{-t}$. Since $(3)^{-t} = (3)^{-1 \cdot t} = (3^{-1})^t = \left(\frac{1}{3}\right)^t$, this formula can also be written $y = 15\left(\frac{1}{3}\right)^t$. $a = 15$ and $b = \frac{1}{3}$. A graph that is both positive and decreasing is the one in (ii).
**(e)** $y = -4(0.98)^t$. So $a = -4$ and $b = 0.98$. Since $a < 0$ and $0 < b < 1$, we want a graph which is both negative and increasing. The graph in (iii) satisfies these conditions.
**(f)** $y = 82(0.8)^{-t}$. Since $(0.8)^{-t} = \left(\frac{8}{10}\right)^{-t} = \left(\frac{8}{10}\right)^{-1 \cdot t} = \left(\left(\frac{8}{10}\right)^{-1}\right)^t = \left(\frac{10}{8}\right)^t = (1.25)^t$ this formula can also be written as $y = 82(1.25)^t$. So $a = 82$ and $b = 1.25$. A graph which is both positive and increasing is the one in (i).

**25. (a)** Since the initial amount is 5.35 and the growth rate is 0.8% per year, the formula is $Q = 5.35(1.008)^t$.
**(b)** At $t = 10$, we have $Q = 5.35(1.008)^{10} = 5.794$.

**29.** The population is growing at a rate of 1.9% per year. So, at the end of each year, the population is $100\% + 1.9\% = 101.9\%$ of what it had been the previous year. The growth factor is 1.019. If $P$ is the population of this country, in millions, and $t$ is the number of years since 2010, then, after one year,

$$P = 70(1.019).$$

$$\text{After two years,} \quad P = 70(1.019)(1.019) = 70(1.019)^2$$

$$\text{After three years,} \quad P = 70(1.019)(1.019)(1.019) = 70(1.019)^3$$

$$\text{After } t \text{ years,} \quad P = 70\underbrace{(1.019)(1.019)\ldots(1.019)}_{t \text{ times}} = 70(1.019)^t$$

**33. (a)** We assume that the price of a movie ticket increases at the rate of 3.5% per year. This means that the price is rising exponentially, so a formula for $p$ is of the form $p = ab^t$. We have $a = 7.50$ and $b = 1 + r = 1.035$. Thus, a formula for $p$ is

$$p = 7.50(1.035)^t.$$

**(b)** In 20 years ($t = 20$) we have $p = 7.50(1.035)^{20} \approx 14.92$. Thus, in 20 years, movie tickets will cost almost \$15 if the inflation rate remains at 3.5%.

**37. (a)** At a time $t$ years after 2005, the population $P$ in millions is $P = 36.8(1.013)^t$.
In 2030, we have $t = 25$, so

$$P = 36.8(1.013)^{25} = 50.826 \text{ million.}$$

Between 2005 and 2030,

$$\text{Increase } = 50.826 - 36.8 = 14.026 \text{ million.}$$

In 2055, we have $t = 50$, so

$$P = 36.8(1.013)^{50} = 70.197 \text{ million.}$$

Between 2030 and 2055,

$$\text{Increase } = 70.197 - 50.826 = 19.371 \text{ million.}$$

**(b)** The increase between 2030 and 2055 is expected to be larger than the increase between 2005 and 2030 because the exponential function is concave up. Both increases are over 25 year periods, but since the graph of the function bends upward, the increase in the later time period is larger.

**41. (a)** In 2012, we have

$$\text{World PV market installations } = 2826(1.62)^5 = 31{,}531.7 \text{ megawatts,}$$

and

$$\text{Japan PV market installations } = 230(0.77)^5 = 62.3 \text{ megawatts.}$$

**(b)** In 2007, the proportion of world market installations in Japan was $230/2826 = 0.081$, or about 8.1%. In 2012, we estimate that the proportion is $62.3/31{,}531.7 = 0.00197$, or less than 0.2%.

**45. (a)** Since $N$ is growing by 5% per year, we know that $N$ is an exponential function of $t$ with growth factor $1 + 0.05 = 1.05$. Since $N = 13.4$ when $t = 0$, we have

$$N = 13.4(1.05)^t.$$

**(b)** In the year 2015, we have $t = 6$ and

$$N = 13.4(1.05)^6 \approx 17.957 \text{ million passengers.}$$

In the year 2005, we have $t = -4$ and

$$N = 13.4(1.05)^{-4} \approx 11.024 \text{ million passengers.}$$

**49.** Let $r$ be the percentage by which the substance decays each year. Every year we multiply the amount of radioactive substance by $1 - r$ to determine the new amount. If $a$ is the amount of the substance on hand originally, we know that after five years, there have been five yearly decreases, by a factor of $1 - r$. Since we know that there will be 60% of $a$, or $0.6a$, remaining after five years (because 40% of the original amount will have decayed), we know that

$$a \cdot \underbrace{(1 - r)^5}_{\substack{\text{five annual decreases} \\ \text{by a factor of } 1 - r}} = 0.6a.$$

Dividing both sides by $a$, we have $(1 - r)^5 = 0.6$, which means that

$$1 - r = (0.6)^{\frac{1}{5}} \approx 0.9029$$

so

$$r \approx 0.09712 = 9.712\%.$$

Each year the substance decays by 9.712%.

**53.** Each time we make a tri-fold, we triple the number of layers of paper, $N(x)$. So $N(x) = 3^x$, where $x$ is the number of folds we make. After 20 folds, the letter would have $3^{20}$ (almost 3.5 billion!) layers. To find out how high our letter would be, we divide the number of layers by the number of sheets in one inch. So the height, $h$, is

$$h = \frac{3^{20} \text{ sheets}}{150 \text{ sheets/inch}} \approx 23{,}245{,}229.34 \text{ inches.}$$

Since there are 12 inches in a foot and 5280 feet in a mile, this gives

$$h \approx 23245229.34 \text{ in} \left( \frac{1 \text{ ft}}{12 \text{ in}} \right) \left( \frac{1 \text{ mile}}{5280 \text{ ft}} \right)$$
$$\approx 366.875 \text{ miles.}$$

**57.** The graph $a_0(b_0)^t$ climbs faster than that of $a_1(b_1)^t$, so $b_0 > b_1$.

**61.** **(a)** We have

$$\text{Total revenue} = \text{No. households} \times \text{Rate per household}$$
$$\text{so} \quad R = N \times r.$$

**(b)** We have

$$\text{Average revenue} = \frac{\text{Total revenue}}{\text{No. students}}$$
$$\text{so} \quad A = \frac{R}{P} = \frac{Nr}{P}.$$

**(c)** We have

$$N_{\text{new}} = N + (2\%)N = 1.02N$$
$$r_{\text{new}} = r + (3\%)r = 1.03r$$

**(d)** We have

$$R_{\text{new}} = N_{\text{new}} \times r_{\text{new}} = (1.02N)(1.03r)$$
$$= 1.0506Nr = 1.0506R.$$

Thus, $R$ increased by 5.06%, or by just over 5%.

(e) We have

$$P_{\text{new}} = P + (8\%)P = 1.08P$$

$$A_{\text{new}} = \frac{R_{\text{new}}}{P_{\text{new}}}$$

$$= \frac{1.0506R}{1.08P} = (0.9728)\left(\frac{R}{P}\right) \approx (97.3\%)A,$$

so the average revenue fell by 2.7%, despite the fact that the tax rate and the tax base both grew.

## Solutions for Section 4.2

### Skill Refresher

**S1.** We have $b^4 \cdot b^6 = b^{4+6} = b^{10}$.

**S5.** We have $f(0) = 5.6(1.043)^0 = 5.6$ and $f(3) = 5.6(1.043)^3 = 6.354$.

**S9.** We have

$$\frac{4}{3}x^5 = 7$$

$$x^5 = 5.25$$

$$x = (5.25)^{1/5} = 1.393.$$

### Exercises

**1. (a)** Since the rate of change is constant, the formula is the linear function $p = 2.50 + 0.03t$.
   **(b)** Since the rate of change is constant, the formula is the linear function $p = 2.50 - 0.07t$.
   **(c)** Since the percent rate of change is constant, the formula is the exponential function $p = 2.50(1.02)^t$.
   **(d)** Since the percent rate of change is constant, the formula is the exponential function $p = 2.50(0.96)^t$.

**5.** We have $g(10) = 50$ and $g(30) = 25$. Using the ratio method, we have

$$\frac{ab^{30}}{ab^{10}} = \frac{g(30)}{g(10)}$$

$$b^{20} = \frac{25}{50}$$

$$b = \left(\frac{25}{50}\right)^{1/20} \approx 0.965936.$$

Now we can solve for $a$:

$$a(0.965936)^{10} = 50$$

$$a = \frac{50}{(0.965936)^{10}} \approx 70.711.$$

so $Q = 70.711(0.966)^t$.

**9.** Let the equation of the exponential curve be $Q = ab^t$. Since this curve passes through the points $(-1, 2)$, $(1, 0.3)$, we have

$$2 = ab^{-1}$$

$$0.3 = ab^1 = ab$$

So,

$$\frac{0.3}{2} = \frac{ab}{ab^{-1}} = b^2,$$

that is, $b^2 = 0.15$, thus $b = \sqrt{0.15} \approx 0.3873$ because $b$ is positive. Since $2 = ab^{-1}$, we have $a = 2b = 2 \cdot 0.3873 = 0.7746$, and the equation of the exponential curve is

$$Q = 0.7746 \cdot (0.3873)^t.$$

13. Since this function is exponential, we know $y = ab^x$. We also know that $(-2, 8/9)$ and $(2, 9/2)$ are on the graph of this function, so

$$\frac{8}{9} = ab^{-2}$$

and

$$\frac{9}{2} = ab^2.$$

From these two equations, we can say that

$$\frac{\frac{9}{2}}{\frac{8}{9}} = \frac{ab^2}{ab^{-2}}.$$

Since $(9/2)/(8/9) = 9/2 \cdot 9/8 = 81/16$, we can re-write this equation to be

$$\frac{81}{16} = b^4.$$

Keeping in mind that $b > 0$, we get

$$b = \sqrt[4]{\frac{81}{16}} = \frac{\sqrt[4]{81}}{\sqrt[4]{16}} = \frac{3}{2}.$$

Substituting $b = 3/2$ in $ab^2 = 9/2$, we get

$$\frac{9}{2} = a\left(\frac{3}{2}\right)^2 = \frac{9}{4}a$$

$$a = \frac{\frac{9}{2}}{\frac{9}{4}} = \frac{9}{2} \cdot \frac{4}{9} = \frac{4}{2} = 2.$$

Thus, $y = 2(3/2)^x$.

17. Let $f$ be the function whose graph is shown. Were $f$ exponential, it would increase by equal factors on equal intervals. However, we see that

$$\frac{f(3)}{f(1)} = \frac{11}{5} = 2.2$$
$$\frac{f(5)}{f(3)} = \frac{30}{11} = 2.7.$$

On these equal intervals, the value of $f$ does not increase by equal factors, so $f$ is not exponential.

21. Note that in the table, the $x$-values are evenly spaced with $\Delta x = 5$.

- Taking ratios, we see that on equally spaced intervals, $f$ appears to decrease by a constant factor:

$$\frac{f(5)}{f(0)} = \frac{85.9}{95.4} = 0.90$$
$$\frac{f(10)}{f(5)} = \frac{77.3}{85.9} = 0.90$$
$$\frac{f(15)}{f(10)} = \frac{69.6}{77.3} = 0.90$$
$$\frac{f(20)}{f(15)} = \frac{62.6}{69.6} = 0.90.$$

This is the hallmark of an exponential function.

- Taking differences, we see that on equally spaced intervals, $h$ appears to decrease by a constant amount:

$$h(5) - h(0) = 36.6 - 37.3 = -0.7$$
$$h(10) - h(5) = 35.9 - 36.6 = -0.7$$
$$h(15) - h(10) = 35.2 - 35.9 = -0.7$$
$$h(20) - h(15) = 34.5 - 35.2 = -0.7.$$

This is the hallmark of a linear function.

- Taking differences, we see that $g$ does not change by a constant factor. For instance,

$$\frac{g(5)}{g(0)} = \frac{40.9}{44.8} = 0.91$$
$$\frac{g(10)}{g(5)} = \frac{36.8}{40.9} = 0.90$$
$$\frac{g(15)}{g(10)} = \frac{32.5}{36.8} = 0.88$$
$$\frac{g(20)}{g(15)} = \frac{28.0}{32.5} = 0.86.$$

Nor, however, does $g$ change by a constant amount:

$$g(5) - g(0) = 40.9 - 44.8 = -3.9$$
$$g(10) - g(5) = 36.8 - 40.9 = -4.1$$
$$g(15) - g(10) = 32.5 - 36.8 = -4.3$$
$$g(20) - g(15) = 28 - 32.5 \quad = -4.5.$$

This means $g$ is neither exponential nor linear.

25. (a) If a function is linear, then the differences in successive function values will be constant. If a function is exponential, the ratios of successive function values will remain constant. Now

$$i(1) - i(0) = 14 - 18 = -4$$

and

$$i(2) - i(1) = 10 - 14 = -4.$$

Checking the rest of the data, we see that the differences remain constant, so $i(x)$ is linear.

(b) We know that $i(x)$ is linear, so it must be of the form

$$i(x) = b + mx,$$

where $m$ is the slope and $b$ is the $y$-intercept. Since at $x = 0$, $i(0) = 18$, we know that the $y$-intercept is 18, so $b = 18$. Also, we know that at $x = 1$, $i(1) = 14$, we have

$$i(1) = b + m \cdot 1$$
$$14 = 18 + m$$
$$m = -4.$$

Thus, $i(x) = 18 - 4x$. The graph of $i(x)$ is shown in Figure 4.1.

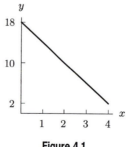

**Figure 4.1**

## Problems

**29.** We have $p = f(x) = ab^x$. From the figure, we see that the starting value is $a = 20$ and that the graph contains the point $(10, 40)$. We have

$$f(10) = 40$$
$$20b^{10} = 40$$
$$b^{10} = 2$$
$$b = 2^{1/10}$$
$$= 1.0718,$$

so $p = 20(1.0718)^x$.

We have $q = g(x) = ab^x$, $g(10) = 40$, and $g(15) = 20$. Using the ratio method, we have

$$\frac{ab^{15}}{ab^{10}} = \frac{g(15)}{g(10)}$$
$$b^5 = \frac{20}{40}$$
$$b = \left(\frac{20}{40}\right)^{1/5}$$
$$= (0.5)^{1/5} = 0.8706.$$

Now we can solve for $a$:

$$a\left((0.5)^{1/5}\right)^{10} = 40$$
$$a = \frac{40}{(0.5)^2}$$
$$= 160.$$

so $q = 160(0.8706)^x$.

**33. (a)** Since this function is exponential, its formula is of the form $f(t) = ab^t$, so

$$f(3) = ab^3$$
$$f(8) = ab^8.$$

From the graph, we know that

$$f(3) = 2000$$
$$f(8) = 5000.$$

So

$$\frac{f(8)}{f(3)} = \frac{ab^8}{ab^3} = \frac{5000}{2000}$$
$$b^5 = \frac{5}{2} = 2.5$$
$$(b^5)^{1/5} = (2.5)^{1/5}$$
$$b = 1.20112.$$

We now know that $f(t) = a(1.20112)^t$. Using either of the pairs of values on the graph, we can find $a$. In this case, we use $f(3) = 2000$. According to the formula,

$$f(3) = a(1.20112)^3$$
$$2000 = a(1.20112)^3$$
$$a = \frac{2000}{(1.20112)^3} \approx 1154.160.$$

The formula we want is $f(t) = 1154.160(1.20112)^t$ or $P = 1154.160(1.20112)^t$.

**(b)** The initial value of the account occurs when $t = 0$.

$$f(0) = 1154.160(1.20112)^0 = 1154.160(1) = \$1154.16.$$

**(c)** The value of $b$, the growth factor, is related to the growth rate, $r$, by

$$b = 1 + r.$$

We know that $b = 1.20112$, so

$$1.20112 = 1 + r$$
$$0.20112 = r$$

Thus, in percentage terms, the annual interest rate is 20.112%.

**37.** We let $W$ represent the winning time and $t$ represent the number of years since 1994.

**(a)** To find the linear function, we first find the slope:

$$\text{Slope} = \frac{\Delta W}{\Delta t} = \frac{41.94 - 43.45}{12 - 0} = -0.126.$$

The vertical intercept is 43.45 so the linear function is $W = 43.45 - 0.126t$. The predicted winning time in 2018 is $W = 43.45 - 0.126(24) = 40.43$ seconds.

**(b)** The time at $t = 0$ is 43.45, so the exponential function is $W = 43.45(a)^t$. We use the fact that $W = 41.94$ when $t = 12$ to find $a$:

$$41.94 = 43.45(a)^{12}$$
$$0.965247 = a^{12}$$
$$a = (0.965247)^{1/12} = 0.997057.$$

The exponential function is $W = 43.45(0.997057)^t$. The predicted winning time in 2018 is $W = 43.45(0.997057)^{24} = 40.48$ seconds.

**41. (a)** Since the rate of change is constant, the increase is linear.

**(b)** Life expectancy is increasing at a constant rate of 3 months, or 0.25 years, each year. The slope is 0.25. When $t = 9$ we have $L = 78.1$. We use the point-slope form to find the linear function:

$$L - 78.1 = 0.25(t - 9)$$
$$L - 78.1 = 0.25t - 2.25$$
$$L = 0.25t + 75.85.$$

**(c)** When $t = 50$, we have $L = 0.25(50) + 75.85 = 88.35$. It the rate of increase continues, babies born in 2050 will have a life expectancy of 88.35 years.

**45. (a)** Assuming linear growth at 250 per year, the population in 2010 would be

$$18{,}500 + 250 \cdot 10 = 21{,}000.$$

Using the population after one year, we find that the percent rate would be $250/18{,}500 \approx 0.013514 = 1.351\%$ per year, so after 10 years the population would be

$$18{,}500(1.013514)^{10} \approx 21{,}158.$$

The town's growth is poorly modeled by both linear and exponential functions.

**(b)** We do not have enough information to make even an educated guess about a formula.

# Solutions for Section 4.3

## Exercises

**1.** **(a)** See Table 4.1.

**Table 4.1**

| $x$ | $-3$ | $-2$ | $-1$ | 0 | 1 | 2 | 3 |
|---|---|---|---|---|---|---|---|
| $f(x)$ | 1/8 | 1/4 | 1/2 | 1 | 2 | 4 | 8 |

**(b)** For large negative values of $x$, $f(x)$ is close to the $x$-axis. But for large positive values of $x$, $f(x)$ climbs rapidly away from the $x$-axis. As $x$ gets larger, $y$ grows more and more rapidly. See Figure 4.2.

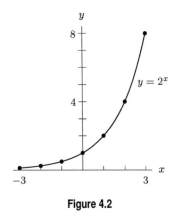

**Figure 4.2**

**5.** Yes, the graphs will cross. The graph of $g(x)$ has a smaller $y$-intercept but increases faster and will eventually overtake the graph of $f(x)$.

**9.** No, the graphs will not cross. Both functions are decreasing but the graph of $f(x)$ has a larger $y$-intercept and is decreasing at a slower rate than $g(x)$ so it will always be above the graph of $g(x)$.

**13.** The function with the smallest $b$ should be the one that is decreasing the fastest. We note that $D$ approaches zero faster than the others, so $D$ has the smallest $b$.

**17.** Graphing $p = 22(0.87)^q$ and tracing along the graph on a calculator gives us an answer of $q = 5.662$. See Figure 4.3.

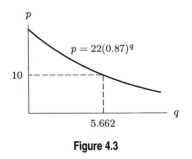

**Figure 4.3**

**21.** As $t$ approaches $\infty$, the value of $ab^t$ approaches zero for any $a$, so the horizontal asymptote is $y = 0$ (the $x$-axis).

## Problems

**25.** A possible graph is shown in Figure 4.4.

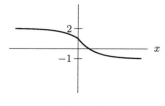

**Figure 4.4**

**29.** It appears in the graph that
  **(a)** $\lim_{x \to -\infty} f(x) = -\infty$
  **(b)** $\lim_{x \to \infty} f(x) = -\infty.$

Of course, we need to be sure that we are seeing all the important features of the graph in order to have confidence in these estimates.

**33.** Increasing: $b > 1, a > 0$ or $0 < b < 1, a < 0$;
Decreasing: $0 < b < 1, a > 0$ or $b > 1, a < 0$;
The function is concave up for $a > 0, 0 < b < 1$ or $b > 1$.

**37. (a)** The growth factor is $1 - 0.0075 = 0.9925$ and the initial value is 651, so we have

$$P = 651(0.9925)^t.$$

  **(b)** Using $t = 10$, we have $P = 651(0.9925)^{10} = 603.790$. If the current trend continues, the population of Baltimore is predicted to be 603,790 in the year 2010.
  **(c)** See Figure 4.5. We see that $t = 22.39$ when $P = 550$. The population is expected to be 550 thousand in approximately the year 2023.

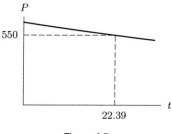

**Figure 4.5**

**41. (a)** For each kilometer above sea level, the atmospheric pressure is $86\% (= 100\% - 14\%)$ of the pressure one kilometer lower. If $P$ represents the number of millibars of pressure and $h$ represents the number of kilometers above sea level. Table 4.2 leads to the formula $P = 1013(0.86)^h$. So, at 50 km, $P = 1013(0.86)^{50} \approx 0.538$ millibars.

**Table 4.2**

| $h$ | $P$ |
|---|---|
| 0 | 1013 |
| 1 | $1013(0.86) = 871.18$ |
| 2 | $871.18(0.86) = 1013(0.86)(0.86) = 1013(0.86)^2$ |
| 3 | $1013(0.86)^2 \cdot (0.86) = 1013(0.86)^3$ |
| 4 | $1013(0.86)^4$ |
| ... | ... |
| $h$ | $1013(0.86)^h$ |

**(b)** If we graph the function $P = 1013(0.86)^h$, we can find the value of $h$ for which $P = 900$. One approach is to see where it intersects the line $P = 900$. Doing so, you will see that at an altitude of $h \approx 0.784$ km, the atmospheric pressure will have dropped to 900 millibars.

**45. (a)** See Figure 4.6. The points appear to represent a function that is increasing and concave up so it makes sense to model these data with an exponential function.

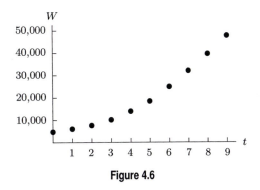

**Figure 4.6**

**(b)** One algorithm used by a calculator or computer gives the exponential regression function as

$$W = 4710(1.306)^t$$

Other algorithms may give different formulas.

**(c)** Since the base of this exponential function is 1.306, the global wind energy generating capacity was increasing at a rate of about 30.6% per year during this period.

# Solutions for Section 4.4

## Exercises

**1. (a)** Suppose $1 is put in the account. The interest rate per month is $0.08/12$. At the end of a year,

$$\text{Balance} = \left(1 + \frac{0.08}{12}\right)^{12} = \$1.08300,$$

which is 108.3% of the original amount. So the effective annual yield is 8.300%.

(b) With weekly compounding, the interest rate per week is 0.08/52. At the end of a year,

$$\text{Balance} = \left(1 + \frac{0.08}{52}\right)^{52} = \$1.08322,$$

which is 108.322% of the original amount. So the effective annual yield is 8.322%.

(c) Assuming it is not a leap year, the interest rate per day is 0.08/365. At the end of a year

$$\text{Balance} = \left(1 + \frac{0.08}{365}\right)^{365} = \$1.08328,$$

which is 108.328% of the original amount. So the effective annual yield is 8.328%.

**5.** (a) The nominal interest rate is 8%, so the interest rate per month is 0.08/12. Therefore, at the end of 3 years, or 36 months,

$$\text{Balance} = \$1000 \left(1 + \frac{0.08}{12}\right)^{36} = \$1270.24.$$

(b) There are 52 weeks in a year, so the interest rate per week is 0.08/52. At the end of $52 \times 3 = 156$ weeks,

$$\text{Balance} = \$1000 \left(1 + \frac{0.08}{52}\right)^{156} = \$1271.01.$$

(c) Assuming no leap years, the interest rate per day is 0.08/365. At the end of $3 \times 365$ days

$$\text{Balance} = \$1000 \left(1 + \frac{0.08}{365}\right)^{3 \cdot 365} = \$1271.22.$$

**9.** (a) If the interest is compounded annually, there will be $\$500 \cdot 1.05 = \$525$ after one year.

(b) If the interest is compounded weekly, there will be $500 \cdot (1 + 0.05/52)^{52} = \$525.62$ after one year.

(c) If the interest is compounded every minute, there will be $500 \cdot (1 + 0.05/525,600)^{525,600} = \$525.64$ after one year.

**13.** (a) The nominal rate is the stated annual interest without compounding, thus 3%.

The effective annual rate for an account paying 1% compounded annually is 3%.

(b) The nominal rate is the stated annual interest without compounding, thus 3%.

With quarterly compounding, there are four interest payments per year, each of which is $3/4 = 0.75\%$. Over the course of the year, this occurs four times, giving an effective annual rate of $1.0075^4 = 1.03034$, which is 3.034%.

(c) The nominal rate is the stated annual interest without compounding, thus 3%.

With daily compounding, there are 365 interest payments per year, each of which is $(3/365)\%$. Over the course of the year, this occurs 365 times, giving an effective annual rate of $(1 + 0.03/365)^{365} = 1.03045$, which is 3.045%.

## Problems

**17.** Let $b$ represent the growth factor, since the investment decreases, $b < 1$. If we start with an investment of $P_0$, then after 12 years, there will be $P_0 b^{12}$ left. But we know that since the investment has decreased by 60% there will be 40% remaining after 12 years. Therefore,

$$P_0 b^{12} = P_0 0.40$$
$$b^{12} = 0.40$$
$$b = (0.40)^{1/12} = 0.92648.$$

This tells us that the value of the investment will be 92.648% of its value the previous year, or that the value of the investment decreases by approximately 7.352% each year, assuming a constant percent decay rate.

**21.** (i) Equation (b). Since the growth factor is 1.12, or 112%, the annual interest rate is 12%.

(ii) Equation (a). An account earning at least 1% monthly will have a monthly growth factor of at least 1.01, which means that the annual (12-month) growth factor will be at least

$$(1.01)^{12} \approx 1.1268.$$

Thus, an account earning at least 1% monthly will earn at least 12.68% yearly. The only account that earns this much interest is account (a).

(iii) Equation (c). An account earning 12% annually compounded semi-annually will earn 6% twice yearly. In $t$ years, there are $2t$ half-years.

(iv) Equations (b), (c) and (d). An account that earns 3% each quarter ends up with a yearly growth factor of $(1.03)^4 \approx$ 1.1255. This corresponds to an annual percentage rate of 12.55%. Accounts (b), (c) and (d) earn less than this. Check this by determining the growth factor in each case.

(v) Equations (a) and (e). An account that earns 6% every 6 months will have a growth factor, after 1 year, of $(1 + 0.06)^2 = 1.1236$, which is equivalent to a 12.36% annual interest rate, compounded annually. Account (a), earning 20% each year, clearly earns more than 6% twice each year, or 12.36% annually. Account (e), which earns 3% each quarter, earns $(1.03)^2 = 1.0609$, or 6.09% every 6 months, which is greater than 6%.

# Solutions for Section 4.5

## Skill Refresher

**S1.** We have $e^{0.07} = 1.073$.

**S5.** We have $f(0) = 2.3e^{0.3(0)} = 2.3$ and $f(4) = 2.3e^{0.3(4)} = 7.636$.

**S9.** Writing the function as
$$f(t) = \left(3e^{0.04t}\right)^3 = 3^3 e^{0.04t \cdot 3} = 27e^{0.12t},$$
we have $a = 27$ and $k = 0.12$.

**S13.** Writing the function as
$$m(x) = \frac{7e^{0.2x}}{\sqrt{3e^x}} = \frac{7}{\sqrt{3}}e^{0.2x} \cdot e^{-0.5x} = \frac{7}{\sqrt{3}}e^{-0.3x},$$
we have $a = \frac{7}{\sqrt{3}}$ and $k = -0.3$.

## Exercises

1. Using the formula $y = ab^x$, each of the functions has the same value for $b$, but different values for $a$ and thus different $y$-intercepts.

   When $x = 0$, the $y$-intercept for $y = e^x$ is 1 since $e^0 = 1$.

   When $x = 0$, the $y$-intercept for $y = 2e^x$ is 2 since $e^0 = 1$ and $2(1) = 2$.

   When $x = 0$, the $y$-intercept for $y = 3e^x$ is 3 since $e^0 = 1$ and $3(1) = 3$.

   Therefore, $y = e^x$ is the bottom graph, above it is $y = 2e^x$ and the top graph is $y = 3e^x$.

5. $y = e^x$ is an increasing exponential function, since $e > 1$. Therefore, it rises when read from left to right. It matches $g(x)$.

   If we rewrite the function $y = e^{-x}$ as $y = (e^{-1})^x$, we can see that in the formula $y = ab^x$, we have $a = 1$ and $b = e^{-1}$. Since $0 < e^{-1} < 1$, this graph has a positive $y$-intercept and falls when read from left to right. Thus its graph is $f(x)$.

   In the function $y = -e^x$, we have $a = -1$. Thus, the vertical intercept is $y = -1$. The graph of $h(x)$ has a negative $y$-intercept.

## Problems

9. $\displaystyle\lim_{x \to \infty} e^{-3x} = 0$.

13. The values of $a$ and $k$ are both positive.

17. **(a)** We see that $Q_0 = 0.01$
    **(b)** Since the exponent is negative, the quantity is decreasing.
    **(c)** The decay rate is 20% per unit time.
    **(d)** Yes, the growth rate is continuous.

**21. (a)** Since the growth rate is not continuous, we have $Q = 8(1.12)^t$. At $t = 10$ we have $Q = 8(1.12)^{10} = 24.847$.

  **(b)** Since the growth rate is continuous, we have $Q = 8e^{0.12t}$. At $t = 10$ we have $Q = 8e^{0.12(10)} = 26.561$. As we expect, the results are similar for continuous and not continuous assumptions, but slightly larger if we assume a continuous growth rate.

**25.** If the population is growing or shrinking at a constant rate of $m$ people per year, the formula is linear. Since the vertical intercept is 3000, we have $P = 3000 + mt$.

If the population is growing or shrinking at a constant percent rate of $r$ percent per year, the formula is exponential in the form $P = a(1 + r)^t$. Since the vertical intercept is 3000, we have $P = 3000(1 + r)^t$.

If the population is growing or shrinking at a constant continuous percent rate of $k$ percent per year, the formula is exponential in the form $P = ae^{kt}$. Since the vertical intercept is 3000, we have $P = 3000e^{kt}$.

We have:

  **(a)** $P = 3000 + 200t$.
  **(b)** $P = 3000(1.06)^t$.
  **(c)** $P = 3000e^{0.06t}$.
  **(d)** $P = 3000 - 50t$.
  **(e)** $P = 3000(0.96)^t$.
  **(f)** $P = 3000e^{-0.04t}$.

**29.**
  • All three investments begin (in year $t = 0$) with \$1000.
  • The investment $V = 1000e^{0.115t}$ earns interest at a continuous annual rate of 11.5%.
  • The investment $V = 1000 \cdot 2^{t/6}$ doubles in value every 6 years.
  • The investment $V = 1000(1.122)^t$ grows by 12.2% every year.

**33. (a)** At the end of 100 years,

$$B = 1200e^{0.03(100)} = 24{,}102.64 \text{ dollars}.$$

  **(b)** Tracing along a graph of $B = 1200e^{0.03t}$ until $B = 50000$ gives $t \approx 124.323$ years.

**37.** The value of the deposit is given by

$$V = 1000e^{0.05t}.$$

To find the effective annual rate, we use the fact that $e^{0.05t} = (e^{0.05})^t$ to rewrite the function as

$$V = 1000(e^{0.05})^t.$$

Since $e^{0.05} = 1.05127$, we have

$$V = 1000(1.05127)^t.$$

This tells us that the effective annual rate is 5.127%.

**41.** To see which investment is best after 1 year, we compute the effective annual yield:

For Bank A, $P = P_0(1 + \frac{0.07}{365})^{365(1)} \approx 1.0725P_0$

For Bank B, $P = P_0(1 + \frac{0.071}{12})^{12(1)} \approx 1.0734P_0$

For Bank C, $P = P_0(e^{0.0705(1)}) \approx 1.0730P_0$

Therefore, the best investment is with Bank B, followed by Bank C and then Bank A.

**45.** The balance in the first bank is $10{,}000(1.05)^8 = \$14{,}774.55$. The balance in the second bank is $10{,}000e^{0.05(8)} = \$14{,}918.25$. The bank with continuously compounded interest has a balance \$143.70 higher.

**49. (a)** The substance decays according to the formula

$$A = 50e^{-0.14t}.$$

  **(b)** At $t = 10$, we have $A = 50e^{-0.14(10)} = 12.330$ mg.
  **(c)** We see in Figure 4.7 that $A = 5$ at approximately $t = 16.45$, which corresponds to the year 2025.

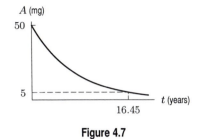

**Figure 4.7**

**53. (a)** The sum is 2.708333333.

**(b)** The sum of $1 + \frac{1}{1} + \frac{1}{1\cdot 2} + \frac{1}{1\cdot 2\cdot 3} + \frac{1}{1\cdot 2\cdot 3\cdot 4} + \frac{1}{1\cdot 2\cdot 3\cdot 4\cdot 5} + \frac{1}{1\cdot 2\cdot 3\cdot 4\cdot 5\cdot 6}$ is 2.718055556.

**(c)** 2.718281828 is the calculator's internal value for $e$. The sum of the first five terms has two digits correct, while the sum of the first seven terms has four digits correct.

**(d)** One approach to finding the number of terms needed to approximate $e$ is to keep a running sum. We already have the total for seven terms displayed, so we can add the eighth term, $\frac{1}{1\cdot 2\cdot 3\cdot 4\cdot 5\cdot 6\cdot 7}$, and compare the result with 2.718281828. Repeat this process until you get the required degree of accuracy. Using this process, we discover that 13 terms are required.

# Solutions for Chapter 4 Review

## Exercises

**1.** For a 10% increase, we multiply by 1.10 to obtain $500 \cdot 1.10 = 550$.

**5.** For a 42% increase, we multiply by 1.42 to obtain $500 \cdot 1.42 = 710$. For a 42% decrease, we multiply by $1 - 0.42 = 0.58$ to obtain $710 \cdot 0.58 = 411.8$.

**9.** The percent of change is given by

$$\text{Percent of change} = \frac{\text{Amount of change}}{\text{Old amount}} \cdot 100\%.$$

So in these two cases,

$$\text{Percent of change from 10 to 12} = \frac{12 - 10}{10} \cdot 100\% = 20\%$$

$$\text{Percent of change from 100 to 102} = \frac{102 - 100}{100} \cdot 100\% = 2\%$$

**13.** This cannot be linear, since $\Delta f(x)/\Delta x$ is not constant, nor can it be exponential, since between $x = 15$ and $x = 12$, we see that $f(x)$ doubles while $\Delta x = 3$. Between $x = 15$ and $x = 16$, we see that $f(x)$ doubles while $\Delta x = 1$, so the percentage increase is not constant. Thus, the function is neither.

## Problems

**17. (a)** $B = B_0(1.042)^1 = B_0(1.042)$, so the effective annual rate is 4.2%.

**(b)** $B = B_0\left(1 + \frac{.042}{12}\right)^{12} \approx B_0(1.0428)$, so the effective annual rate is approximately 4.28%.

**(c)** $B = B_0 e^{0.042(1)} \approx B_0(1.0429)$, so the effective annual rate is approximately 4.29%.

**21.** Since $g(x) = ab^x$, we can say that $g(\frac{1}{2}) = ab^{1/2}$ and $g(\frac{1}{4}) = ab^{1/4}$. Since we know that $g(\frac{1}{2}) = 4$ and $g(\frac{1}{4}) = 2\sqrt{2}$, we can conclude that

$$ab^{1/2} = 4 = 2^2$$

and

$$ab^{1/4} = 2\sqrt{2} = 2 \cdot 2^{1/2} = 2^{3/2}.$$

Forming ratios, we have

$$\frac{ab^{1/2}}{ab^{1/4}} = \frac{2^2}{2^{3/2}}$$
$$b^{1/4} = 2^{1/2}$$
$$(b^{1/4})^4 = (2^{1/2})^4$$
$$b = 2^2 = 4.$$

Now we know that $g(x) = a(4)^x$, so $g(\frac{1}{2}) = a(4)^{1/2} = 2a$. Since we also know that $g(\frac{1}{2}) = 4$, we can say

$$2a = 4$$
$$a = 2.$$

Therefore $g(x) = 2(4)^x$.

**25. (a)** If $f$ is linear, then $f(x) = b + mx$, where $m$, the slope, is given by:

$$m = \frac{\Delta y}{\Delta x} = \frac{f(2) - f(-3)}{(2) - (-3)} = \frac{20 - \frac{5}{8}}{5} = \frac{\frac{155}{8}}{5} = \frac{31}{8}.$$

Using the fact that $f(2) = 20$, and substituting the known values for $m$, we write

$$20 = b + m(2)$$
$$20 = b + \left(\frac{31}{8}\right)(2)$$
$$20 = b + \frac{31}{4}$$

which gives

$$b = 20 - \frac{31}{4} = \frac{49}{4}.$$

So, $f(x) = \frac{31}{8}x + \frac{49}{4}$.

**(b)** If $f$ is exponential, then $f(x) = ab^x$. We know that $f(2) = ab^2$ and $f(2) = 20$. We also know that $f(-3) = ab^{-3}$ and $f(-3) = \frac{5}{8}$. So

$$\frac{f(2)}{f(-3)} = \frac{ab^2}{ab^{-3}} = \frac{20}{\frac{5}{8}}$$

$$b^5 = 20 \times \frac{8}{5} = 32$$
$$b = 2.$$

Thus, $f(x) = a(2)^x$. Solve for $a$ by using $f(2) = 20$ and (with $b = 2$), $f(2) = a(2)^2$.

$$20 = a(2)^2$$
$$20 = 4a$$
$$a = 5.$$

Thus, $f(x) = 5(2)^x$.

**29.** The formula is of the form $y = ab^x$. Since the points $(-1, 1/15)$ and $(2, 9/5)$ are on the graph, so

$$\frac{1}{15} = ab^{-1}$$
$$\frac{9}{5} = ab^2.$$

Taking the ratio of the second equation to the first we obtain

$$\frac{9/5}{1/15} = \frac{ab^2}{ab^{-1}}$$
$$27 = b^3$$
$$b = 3.$$

Substituting this value of $b$ into $\frac{1}{15} = ab^{-1}$ gives

$$\frac{1}{15} = a(3)^{-1}$$
$$\frac{1}{15} = \frac{1}{3}a$$
$$a = \frac{1}{15} \cdot 3$$
$$a = \frac{1}{5}.$$

Therefore $y = \frac{1}{5}(3)^x$ is a possible formula for this function.

**33. (a)** If $P$ is linear, then $P(t) = b + mt$ and

$$m = \frac{\Delta P}{\Delta t} = \frac{P(13) - P(7)}{13 - 7} = \frac{3.75 - 3.21}{13 - 7} = \frac{0.54}{6} = 0.09.$$

So $P(t) = b + 0.09t$ and $P(7) = b + 0.09(7)$. We can use this and the fact that $P(7) = 3.21$ to say that

$$3.21 = b + 0.09(7)$$
$$3.21 = b + 0.63$$
$$2.58 = b.$$

So $P(t) = 2.58 + 0.09t$. The slope is 0.09 million people per year. This tells us that, if its growth is linear, the country grows by $0.09(1,000,000) = 90,000$ people every year.

**(b)** If $P$ is exponential, $P(t) = ab^t$. So

$$P(7) = ab^7 = 3.21$$

and

$$P(13) = ab^{13} = 3.75.$$

We can say that

$$\frac{P(13)}{P(7)} = \frac{ab^{13}}{ab^7} = \frac{3.75}{3.21}$$
$$b^6 = \frac{3.75}{3.21}$$
$$(b^6)^{1/6} = \left(\frac{3.75}{3.21}\right)^{1/6}$$
$$b = 1.026.$$

Thus, $P(t) = a(1.026)^t$. To find $a$, note that

$$P(7) = a(1.026)^7 = 3.21$$
$$a = \frac{3.21}{(1.026)^7} = 2.68.$$

We have $P(t) = 2.68(1.026)^t$. Since $b = 1.026$ is the growth factor, the country's population grows by about 2.6% per year, assuming exponential growth.

**37.** We have $\lim\limits_{x \to -\infty} (15 - 5e^{3x}) = 15 - 5 \cdot 0 = 15.$

**41.** Since $N = 10$ when $t = 0$, we use $N = 10b^t$ for some base $b$. Since $N = 20000$ when $t = 62$, we have

$$N = 10b^t$$
$$20000 = 10b^{62}$$
$$b^{62} = \frac{20000}{10} = 2000$$
$$b = (2000)^{1/62} = 1.13.$$

An exponential formula for the brown tree snake population is

$$N = 10(1.13)^t.$$

The population has been growing by about 13% per year.

**45.** A possible graph is shown in Figure 4.8.

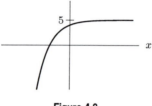

**Figure 4.8**

**49.** According to Figure 4.9, $f$ seems to approach its horizontal asymptote, $y = 0$, faster. To convince yourself, compare values of $f$ and $g$ for very large values of $x$.

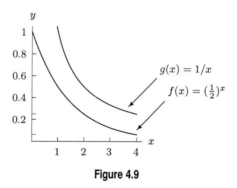

**Figure 4.9**

**53. (a)** This population initially numbers 5200, and it grows in size by 11.8% every year.
**(b)** This population initially numbers 4600. There are $12t$ months in $t$ years, which means the population grows by a factor 1.01 twelve times each year, or once every month. In other words, this population grows by 1% every month.
**(c)** This population initially numbers 3800. It decreases by one-half (50%) every twelve years.
**(d)** This population initially numbers 8000. It grows at a continuous annual rate of 7.78%.
**(e)** Note that unlike the other functions, this is a linear function in point-slope form. It tells us that this population numbers 1675 in year $t = 30$, and the it falls by 25 members every year.

**57.** Writing the function as

$$p(x) = \frac{7e^{6x} \cdot \sqrt{e} \cdot (2e^x)^{-1}}{10e^{4x}} = \frac{7 \cdot \sqrt{e} \cdot 2^{-1}}{10} e^{(6x - x - 4x)} = \frac{7\sqrt{e}}{20} e^x,$$

we have $a = \frac{7\sqrt{e}}{20}$ and $k = 1$.

**61.** We have $g(n) = \sqrt{f(n) \cdot f(n-1)}$ where

$$f(n) = 1000 \cdot 2^{-\frac{1}{4} - \frac{n}{2}}$$
$$= 1000 \cdot 2^{-\frac{1}{4}} \cdot 2^{-\frac{n}{2}}$$
$$\text{and} \quad f(n-1) = 1000 \cdot 2^{-\frac{1}{4} - \frac{n-1}{2}}$$
$$= 1000 \cdot 2^{-\frac{1}{4}} \cdot 2^{-\frac{n-1}{2}}$$
$$= 1000 \cdot 2^{-\frac{1}{4}} \cdot 2^{\frac{1-n}{2}}$$
$$= 1000 \cdot 2^{-\frac{1}{4}} \cdot 2^{\frac{1}{2} - \frac{n}{2}}$$
$$= 1000 \cdot 2^{-\frac{1}{4}} \cdot 2^{\frac{1}{2}} \cdot 2^{-\frac{n}{2}}$$
$$= 1000 \cdot 2^{-\frac{1}{4} + \frac{1}{2}} \cdot 2^{-\frac{n}{2}}$$
$$= 1000 \cdot 2^{\frac{1}{4}} \cdot 2^{-\frac{n}{2}}$$
$$\text{so} \quad f(n) \cdot f(n-1) = \left(1000 \cdot 2^{-\frac{1}{4}} 2^{-\frac{n}{2}}\right) \left(1000 \cdot 2^{\frac{1}{4}} \cdot 2^{-\frac{n}{2}}\right)$$
$$= 1000 \cdot 1000 \cdot 2^{-\frac{1}{4}} \cdot 2^{\frac{1}{4}} \cdot 2^{-\frac{n}{2}} \cdot 2^{-\frac{n}{2}}$$
$$= 1{,}000{,}000 \cdot 2^{-n}.$$

This means

$$g(n) = \sqrt{f(n) \cdot f(n-1)}$$
$$= \sqrt{1{,}000{,}000 \cdot 2^{-n}}$$
$$= \sqrt{1{,}000{,}000} \sqrt{\cdot 2^{-n}}$$
$$= 1000 \left(2^{-n}\right)^{0.5}$$
$$= 1000 \cdot 2^{-0.5n}$$
$$= 1000 \left(2^{-0.5}\right)^{n}$$
$$= 1000(0.7071)^{n}.$$

**65.** In (a), we see that the graph of $g$ starts out below the graph of $f$. In (c), we see that at some point, the lower graph rises to intersect the higher graph, which tells us that $g$ grows faster than $f$. This means that $d > b$.

**69. (a)**

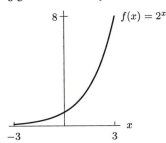

**(b)** The point $(0, 1)$ is on the graph. So is $(0.01, 1.00696)$. Taking $\frac{y_2 - y_1}{x_2 - x_1}$, we get an estimate for the slope of 0.696. We may zoom in still further to find that $(0.001, 1.000693)$ is on the graph. Using this and the point $(0, 1)$ we would get a slope of 0.693. Zooming in still further we find that the slope stabilizes at around 0.693; so, to two digits of accuracy, the slope is 0.69.

**(c)** Using the same method as in part (b), we find that the slope is $\approx 1.10$.

**(d)** We might suppose that the slope of the tangent line at $x = 0$ increases as $b$ increases. Trying a few values, we see that this is the case. Then we can find the correct $b$ by trial and error: $b = 2.5$ has slope around 0.916, $b = 3$ has slope around 1.1, so $2.5 < b < 3$. Trying $b = 2.75$ we get a slope of 1.011, just a little too high. $b = 2.7$ gives a slope of 0.993, just a little too low. $b = 2.72$ gives a slope of 1.0006, which is as good as we can do by giving $b$ to two decimal places. Thus $b \approx 2.72$.

In fact, the slope is exactly 1 when $b = e = 2.718 \ldots$.

**73.** At $x = 50$,

$$y = 5000e^{-50/40} = 1432.5240.$$

At $x = 150$,

$$y = 5000e^{-150/40} = 117.5887.$$

We have $q(50) = 1432.524$ and $q(150) = 117.5887$. This gives $y = b + mx$ where

$$m = \frac{q(150) - q(50)}{150 - 50} = \frac{117.5887 - 1432.524}{100} = -13.1.$$

Solving for $b$, we have

$$q(50) = b - 13.1(50)$$
$$b = q(50) + 13.1(50)$$
$$= 1432.524 + 13.1(50)$$
$$= 2090,$$

so $y = -13.1x + 2090$.

**77. (a)** The data points are approximately as shown in Table 4.3. This results in $a \approx 15.269$ and $b \approx 1.122$, so $E(t) = 15.269(1.122)^t$.

**Table 4.3**

| $t$ (years) | 0 | 1 | 2 | 3 | 4 | 5 | 6 | 7 | 8 | 9 | 10 | 11 | 12 |
|---|---|---|---|---|---|---|---|---|---|---|---|---|---|
| $E(t)$ (thousands) | 22 | 18 | 20 | 20 | 22 | 22 | 19 | 30 | 45 | 42 | 62 | 60 | 65 |

**(b)** In 1997 we have $t = 17$ so $E(17) = 15.269(1.122)^{17} \approx 108,066$.

**(c)** The model is probably not a good predictor of emigration in the year 2010 because Hong Kong was transferred to Chinese rule in 1997. Thus, conditions which affect emigration in 2010 may be markedly different than they were in the period from 1989 to 1992, for which data is given. In 2000, emigration was about 12,000.

**81. (a)** See Figure 4.10.

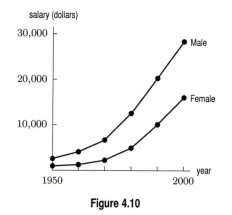

**Figure 4.10**

**(b)** For females, $W(1950) = ae^{b(0)} = a = 953$. Using trial and error, we find a value for $b = 0.062$ which approximates values in the table. So a possible formula for the median income of women is $W_F(t) = 953e^{0.062(t-1950)}$.

For males, $W(1950) = ae^{b(0)} = a = 2570$. A possible value for $b$ is 0.051. These values give us the formula for median income of men of $W_M(t) = 2570e^{0.051(t-1950)}$.

(c) Through the year 2000, women's incomes trail behind those of men. See Figure 4.11. However Figure 4.12 shows women's incomes eventually overtake men's.

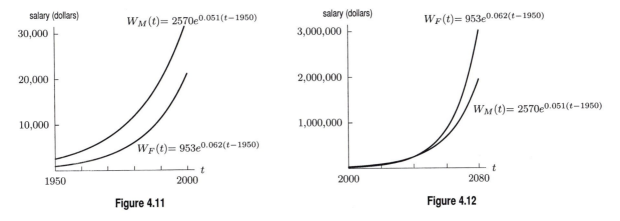

Figure 4.11                                    Figure 4.12

(d) The graph over a larger interval predicts a more promising outlook for equality in incomes. If we look carefully at each formula, $W_F(t) = 953e^{0.062(t-1950)}$ and $W_M(t) = 2570e^{0.051(t-1950)}$, we observe that women should ultimately catch up with, and even surpass, the income of men. The justification for this is the fact that the exponent in $W_F$ is larger than the exponent in $W_M$, so $W_F$ increases more quickly than $W_M$. See Figure 4.12.

   The graph suggests that women will earn approximately the same amount as men in about the year 2060.

(e) The trends observed may not continue into the future, so the model may not apply. Thus these predictions are not reliable.

## CHECK YOUR UNDERSTANDING

1. True. If the constant rate is $r$ then the formula is $f(t) = a \cdot (1 + r)^t$. The function decreases when $0 < 1 + r < 1$ and increases when $1 + r > 1$.

5. False. The annual growth factor would be 1.04, so $S = S_0(1.04)^t$.

9. True. The initial value means the value of $Q$ when $t = 0$, so $Q = f(0) = a \cdot b^0 = a \cdot 1 = a$.

13. False. This is the formula of a linear function.

17. True. The irrational number $e = 2.71828 \cdots$ has this as a good approximation.

21. True. The initial value is 200 and the growth factor is 1.04.

25. True. Since $k$ is the continuous growth rate and negative, $Q$ is decreasing.

29. True. The interest from any quarter is compounded in subsequent quarters.

## Solutions to Skills for Chapter 4

1. $(-5)^2 = (-5)(-5) = 25$

5. $\frac{5^3}{5^2} = 5^{3-2} = 5^1 = 5$

9. $\sqrt{4^2} = 4$

13. Since the base of 2 is the same in both numerator and denominator, we have $\frac{2^7}{2^3} = 2^{7-3} = 2^4$ or $2 \cdot 2 \cdot 2 \cdot 2$ or 16.

17. The order of operations tells us to find $10^3$ and then multiply by 2.1. Therefore $(2.1)\left(10^3\right) = (2.1)(1,000) = 2,100$.

**21.** $16^{5/4} = (2^4)^{5/4} = 2^5 = 32$

**25.** Exponentiation is done first, with the result that $(-1)^3 = -1$. Therefore $(-1)^3\sqrt{36} = (-1)\sqrt{36} = (-1)(6) = -6$.

**29.** $3^{-3/2} = \frac{1}{3^{3/2}} = \frac{1}{(3^3)^{1/2}} = \frac{1}{(27)^{1/2}} = \frac{1}{(9 \cdot 3)^{1/2}} = \frac{1}{9^{1/2} \cdot 3^{1/2}} = \frac{1}{3\sqrt{3}}$

**33.** The cube root of 0.125 is 0.5. Therefore $(0.125)^{1/3} = \sqrt[3]{0.125} = 0.5$.

**37.** $\sqrt{x^5 y^4} = (x^5 \cdot y^4)^{1/2} = x^{5/2} \cdot y^{4/2} = x^{5/2} y^2$

**41.** $\sqrt{r^3} = (r^3)^{1/2} = r^{3/2}$

**45.**
$$\begin{aligned}
\sqrt{48 u^{10} v^{12} y^5} &= (48)^{1/2} \cdot (u^{10})^{1/2} \cdot (v^{12})^{1/2} \cdot (y^5)^{1/2} \\
&= (16 \cdot 3)^{1/2} u^5 v^6 y^{5/2} \\
&= 16^{1/2} \cdot 3^{1/2} \cdot u^5 v^6 y^{5/2} \\
&= 4\sqrt{3} u^5 v^6 y^{5/2}
\end{aligned}$$

**49.**
$$(3AB)^{-1} \left(A^2 B^{-1}\right)^2 = \left(3^{-1} \cdot A^{-1} \cdot B^{-1}\right)\left(A^4 \cdot B^{-2}\right) = \frac{A^4}{3^1 \cdot A^1 \cdot B^1 \cdot B^2} = \frac{A^3}{3B^3}.$$

**53.** $\dfrac{a^{n+1} 3^{n+1}}{a^n 3^n} = a^{n+1-n} 3^{n+1-n} = a^1 \cdot 3^1 = 3a$

**57.** $-32^{3/5} = -(\sqrt[5]{32})^3 = -(2)^3 = -8$

**61.** $64^{-3/2} = (\sqrt{64})^{-3} = (8)^{-3} = \left(\dfrac{1}{8}\right)^3 = \dfrac{1}{512}$

**65.** We have
$$\begin{aligned}
7x^4 &= 20x^2 \\
\frac{x^4}{x^2} &= \frac{20}{7} \\
x^2 &= 20/7 \\
x &= \pm(20/7)^{1/2} = \pm 1.690.
\end{aligned}$$

**69.** False

**73.** True

**77.** We have
$$\begin{aligned}
5^x &= 32 \\
(2^a)^x &= 2^5 \\
2^{ax} &= 2^5 \\
ax &= 5 \\
x &= \frac{5}{a}.
\end{aligned}$$

**81.** We have
$$\begin{aligned}
5^x &= 7 \\
(2^a)^x &= 2^b \\
2^{ax} &= 2^b \\
ax &= b \\
x &= \frac{b}{a}.
\end{aligned}$$

# CHAPTER FIVE

## Solutions for Section 5.1

### Skill Refresher

**S1.** Since $1{,}000{,}000 = 10^6$, we have $x = 6$.

**S5.** Since $e^w$ can never equal zero, the equation $e^w = 0$ has no solution. It similarly follows for any constant $b$, $b^x$ can never equal zero.

**S9.** Since $\sqrt[4]{0.1} = 0.1^{1/4} = \left(10^{-1}\right)^{1/4} = 10^{-1/4}$, we have $2t = -\dfrac{1}{4}$. Solving for $t$ we therefore have $t = -\dfrac{1}{8}$.

### Exercises

**1.** The statement is equivalent to $19 = 10^{1.279}$.

**5.** The statement is equivalent to $P = 10^t$.

**9.** The statement is equivalent to $v = \log \alpha$.

**13.** We are solving for an exponent, so we use logarithms. We can use either the common logarithm or the natural logarithm. Since $2^3 = 8$ and $2^4 = 16$, we know that $x$ must be between 3 and 4. Using the log rules, we have

$$2^x = 11$$
$$\log(2^x) = \log(11)$$
$$x \log(2) = \log(11)$$
$$x = \frac{\log(11)}{\log(2)} = 3.459.$$

If we had used the natural logarithm, we would have

$$x = \frac{\ln(11)}{\ln(2)} = 3.459.$$

**17.** We begin by dividing both sides by 17 to isolate the exponent:

$$\frac{48}{17} = (2.3)^w.$$

We then take the log of both sides and use the rules of logs to solve for $w$:

$$\log \frac{48}{17} = \log(2.3)^w$$
$$\log \frac{48}{17} = w \log(2.3)$$
$$w = \frac{\log \frac{48}{17}}{\log(2.3)} = 1.246.$$

### Problems

**21. (a)**    $\log(10 \cdot 100) = \log 1000 = 3$
$\log 10 + \log 100 = 1 + 2 = 3$

**(b)**    $\log(100 \cdot 1000) = \log 100,000 = 5$

$\log 100 + \log 1000 = 2 + 3 = 5$

**(c)**    $\log \dfrac{10}{100} = \log \dfrac{1}{10} = \log 10^{-1} = -1$

$\log 10 - \log 100 = 1 - 2 = -1$

**(d)**    $\log \dfrac{100}{1000} = \log \dfrac{1}{10} = \log 10^{-1} = -1$

$\log 100 - \log 1000 = 2 - 3 = -1$

**(e)**  $\log 10^2 = 2$

$2 \log 10 = 2(1) = 2$

**(f)**  $\log 10^3 = 3$

$3 \log 10 = 3(1) = 3$

In each case, both answers are equal. This reflects the properties of logarithms.

**25.** Using properties of logs, we have

$$\ln(25(1.05)^x) = 6$$
$$\ln(25) + x\ln(1.05) = 6$$
$$x\ln(1.05) = 6 - \ln(25)$$
$$x = \frac{6 - \ln(25)}{\ln(1.05)} = 57.002.$$

**29.** Using properties of logs, we have

$$\log(5x^3) = 2$$
$$\log 5 + 3\log x = 2$$
$$3\log x = 2 - \log 5$$
$$\log x = \frac{2 - \log 5}{3}$$
$$x = 10^{(2 - \log 5)/3} = 2.714.$$

Notice that to solve for $x$, we had to convert from an equation involving logs to an equation involving exponents in the last step.

An alternate way to solve the original equation is to begin by converting from an equation involving logs to an equation involving exponents:

$$\log(5x^3) = 2$$
$$5x^3 = 10^2 = 100$$
$$x^3 = \frac{100}{5} = 20$$
$$x = (20)^{1/3} = 2.714.$$

Of course, we get the same answer with both methods.

**33.** **(a)** $\log 3 = \log \frac{15}{5} = \log 15 - \log 5$

**(b)** $\log 25 = \log 5^2 = 2\log 5$

**(c)** $\log 75 = \log(15 \cdot 5) = \log 15 + \log 5$

**37.**

$$e^{0.044t} = 6$$
$$\ln e^{0.044t} = \ln 6$$
$$0.044t = \ln 6$$
$$t = \frac{\ln 6}{0.044}.$$

**41.**

$$0.4(\frac{1}{3})^{3x} = 7 \cdot 2^{-x}$$

$$0.4(\frac{1}{3})^{3x} \cdot 2^x = 7 \cdot 2^{-x} \cdot 2^x = 7$$

$$0.4 \left((\frac{1}{3})^3\right)^x \cdot 2^x = 7$$

$$\left((\frac{1}{3})^3 \cdot 2\right)^x = \frac{7}{0.4} = 7\left(\frac{5}{2}\right) = \frac{35}{2}$$

$$\log(\frac{2}{27})^x = \log\frac{35}{2}$$

$$x\log(\frac{2}{27}) = \log\frac{35}{2}$$

$$x = \frac{\log(\frac{35}{2})}{\log(\frac{2}{27})}.$$

**45.** Taking natural logs, we get

$$e^{x+4} = 10$$

$$\ln e^{x+4} = \ln 10$$

$$x + 4 = \ln 10$$

$$x = \ln 10 - 4$$

**49.** $\log(2x + 5) \cdot \log(9x^2) = 0$

In order for this product to equal zero, we know that one or both terms must be equal to zero. Thus, we will set each of the factors equal to zero to determine the values of $x$ for which the factors will equal zero. We have

$$\log(2x + 5) = 0 \qquad \text{or} \qquad \log(9x^2) = 0$$

$$2x + 5 = 1 \qquad\qquad\qquad 9x^2 = 1$$

$$2x = -4 \qquad\qquad\qquad x^2 = \frac{1}{9}$$

$$x = -2 \qquad\qquad\qquad x = \frac{1}{3} \text{ or } x = -\frac{1}{3}.$$

Thus our solutions are $x = -2, \frac{1}{3}$, or $-\frac{1}{3}$.

**53.** Doubling $n$, we have

$$\log(2n) = \log 2 + \log n = 0.3010 + \log n.$$

Thus, doubling a quantity increases its log by 0.3010.

**57.** A standard graphing calculator will evaluate both of these expressions to 1. However, by taking logs, we see that

$$\ln A = \ln\left(5^{3^{-47}}\right)$$

$$= 3^{-47}\ln 5$$

$$= 6.0531 \times 10^{-23} \quad \text{with calculator}$$

$$\ln B = \ln\left(7^{5^{-32}}\right)$$

$$= 5^{-32}\ln 7$$

$$= 8.3576 \times 10^{-23} \quad \text{with calculator.}$$

Since $\ln B > \ln A$, we see that $B > A$.

## Solutions for Section 5.2

### Skill Refresher

**S1.** Rewrite as $10^{-\log 5x} = 10^{\log(5x)^{-1}} = (5x)^{-1}$.

**S5.** Taking logs of both sides we get

$$\log(4^x) = \log 9.$$

This gives

$$x \log 4 = \log 9$$

or in other words

$$x = \frac{\log 9}{\log 4} = 1.585.$$

**S9.** We begin by converting to exponential form:

$$\log(2x + 7) = 2$$
$$10^{\log(2x+7)} = 10^2$$
$$2x + 7 = 100$$
$$2x = 93$$
$$x = \frac{93}{2}.$$

### Exercises

**1.** We want $25e^{0.053t} = 25(e^{0.053})^t = ab^t$, so we choose $a = 25$ and $b = e^{0.053} = 1.0544$. The given exponential function is equivalent to the exponential function $y = 25(1.0544)^t$. The annual percent growth rate is 5.44% and the continuous percent growth rate per year is 5.3% per year.

**5.** The continuous percent growth rate is the value of $k$ in the equation $Q = ae^{kt}$, which is 7.

To convert to the form $Q = ab^t$, we first say that the right sides of the two equations equal each other (since each equals $Q$), and then we solve for $a$ and $b$. Thus, we have $ab^t = 4e^{7t}$. At $t = 0$, we can solve for $a$:

$$ab^0 = 4e^{7 \cdot 0}$$
$$a \cdot 1 = 4 \cdot 1$$
$$a = 4.$$

Thus, we have $4b^t = 4e^{7t}$, and we solve for $b$:

$$4b^t = 4e^{7t}$$
$$b^t = e^{7t}$$
$$b^t = \left(e^7\right)^t$$
$$b = e^7 \approx 1096.633.$$

Therefore, the equation is $Q = 4 \cdot 1096.633^t$.

**9.** To convert to the form $Q = ae^{kt}$, we first say that the right sides of the two equations equal each other (since each equals $Q$), and then we solve for $a$ and $k$. Thus, we have $ae^{kt} = 12(0.9)^t$. At $t = 0$, we can solve for $a$:

$$ae^{k \cdot 0} = 12(0.9)^0$$
$$a \cdot 1 = 12 \cdot 1$$
$$a = 12.$$

Thus, we have $12e^{kt} = 12(0.9)^t$, and we solve for $k$:

$$12e^{kt} = 12(0.9)^t$$
$$e^{kt} = (0.9)^t$$
$$\left(e^k\right)^t = (0.9)^t$$
$$e^k = 0.9$$
$$\ln e^k = \ln 0.9$$
$$k = \ln 0.9 \approx -0.105.$$

Therefore, the equation is $Q = 12e^{-0.105t}$.

**13.** We have $a = 230$, $b = 1.182$, $r = b - 1 = 18.2\%$, and $k = \ln b = 0.1672 = 16.72\%$.

**17.** Writing this as $Q = 12.1(10^{-0.11})^t$, we have $a = 12.1$, $b = 10^{-0.11} = 0.7762$, $r = b - 1 = -22.38\%$, and $k = \ln b = -25.32\%$.

## Problems

**21.** Let $t$ be the doubling time, then the population is $2P_0$ at time $t$, so

$$2P_0 = P_0 e^{0.2t}$$
$$2 = e^{0.2t}$$
$$0.2t = \ln 2$$
$$t = \frac{\ln 2}{0.2} \approx 3.466.$$

**25.** The growth factor for Tritium should be $1 - 0.05471 = 0.94529$, since it is decaying by $5.471\%$ per year. Therefore, the decay equation starting with a quantity of $a$ should be:

$$Q = a(0.94529)^t,$$

where $Q$ is quantity remaining and $t$ is time in years. The half life will be the value of $t$ for which $Q$ is $a/2$, or half of the initial quantity $a$. Thus, we solve the equation for $Q = a/2$:

$$\frac{a}{2} = a(0.94529)^t$$
$$\frac{1}{2} = (0.94529)^t$$
$$\log(1/2) = \log(0.94529)^t$$
$$\log(1/2) = t \log(0.94529)$$
$$t = \frac{\log(1/2)}{\log(0.94529)} = 12.320.$$

So the half-life is about 12.3 years.

**29. (a)** We find $k$ so that $ae^{kt} = a(1.08)^t$. We want $k$ so that $e^k = 1.08$. Taking logs of both sides, we have $k = \ln(1.08) = 0.0770$. An interest rate of $7.70\%$, compounded continuously, is equivalent to an interest rate of $8\%$, compounded annually.

**(b)** We find $b$ so that $ab^t = ae^{0.06t}$. We take $b = e^{0.06} = 1.0618$. An interest rate of $6.18\%$, compounded annually, is equivalent to an interest rate of $6\%$, compounded continuously.

**33. (a)** Let $P(t) = P_0 b^t$ describe our population at the end of $t$ years. Since $P_0$ is the initial population, and the population doubles every 15 years, we know that, at the end of 15 years, our population will be $2P_0$. But at the end of 15 years, our population is $P(15) = P_0 b^{15}$. Thus

$$P_0 b^{15} = 2P_0$$

$$b^{15} = 2$$
$$b = 2^{\frac{1}{15}} \approx 1.04729$$

Since $b$ is our growth factor, the population is, yearly, 104.729% of what it had been the previous year. Thus it is growing by 4.729% per year.

**(b)** Writing our formula as $P(t) = P_0 e^{kt}$, we have $P(15) = P_0 e^{15k}$. But we already know that $P(15) = 2P_0$. Therefore,

$$P_0 e^{15k} = 2P_0$$
$$e^{15k} = 2$$
$$\ln e^{15k} = \ln 2$$
$$15k \ln e = \ln 2$$
$$15k = \ln 2$$
$$k = \frac{\ln 2}{15} \approx 0.04621.$$

This tells us that we have a continuous annual growth rate of 4.621%.

**37.** Since $W = 90$ when $t = 0$, we have $W = 90e^{kt}$. We use the doubling time to find $k$:

$$W = 90e^{kt}$$
$$2(90) = 90e^{k(3)}$$
$$2 = e^{3k}$$
$$\ln 2 = 3k$$
$$k = \frac{\ln 2}{3} = 0.231.$$

World wind energy generating capacity is growing at a continuous rate of 23.1% per year. We have $W = 90e^{0.231t}$.

**41. (a)** We see that the initial value of $Q$ is 150 mg and the quantity of caffeine has dropped to half that at $t = 4$. The half-life of caffeine is about 4 hours. Notice that no matter where we start on the graph, the quantity will be halved four hours later.

**(b)** We use $Q = 150e^{kt}$ and the fact that the quantity decays from 150 to 75 in 4 hours. Solving for $k$, we have:

$$150e^{k(4)} = 75$$
$$e^{4k} = 0.5$$
$$4k = \ln(0.5)$$
$$k = \frac{\ln(0.5)}{4} = -0.173.$$

The continuous percent decay rate is $-17.3\%$ per hour. The formula is $Q = 150e^{-0.173t}$.

**45. (a)** Initially, the population is $P = 300 \cdot 2^{0/20} = 300 \cdot 2^0 = 300$. After 20 years, the population reaches $P = 300 \cdot 2^{20/20} = 300 \cdot 2^1 = 600$.

**(b)** To find when the population reaches $P = 1000$, we solve the equation:

$$300 \cdot 2^{t/20} = 1000$$
$$2^{t/20} = \frac{1000}{300} = \frac{10}{3} \qquad \text{dividing by 300}$$
$$\log\left(2^{t/20}\right) = \log\left(\frac{10}{3}\right) \qquad \text{taking logs}$$
$$\left(\frac{t}{20}\right) \cdot \log 2 = \log\left(\frac{10}{3}\right) \qquad \text{using a log property}$$
$$t = \frac{20\log(10/3)}{\log 2} = 34.739,$$

and so it will take the population a bit less than 35 years to reach 1000.

**49.** **(a)** We want a function of the form $R(t) = A B^t$. We know that when $t = 0$, $R(t) = 200$ and when $t = 6$, $R(t) = 100$. Therefore, $100 = 200B^6$ and $B \approx 0.8909$. So $R(t) = 200(0.8909)^t$.

**(b)** Setting $R(t) = 120$ we have $120 = 200(0.8909)^t$ and so $t = \dfrac{\ln 0.6}{\ln 0.8909} \approx 4.422$.

**(c)** Between $t = 0$ and $t = 2$

$$\frac{\Delta R(t)}{\Delta t} = \frac{200(0.8909)^2 - 200(0.8909)^0}{2} = -20.630$$

Between $t = 2$ and $t = 4$

$$\frac{\Delta R(t)}{\Delta t} = \frac{200(0.8909)^4 - 200(0.8909)^2}{2} = -16.374$$

Between $x = 4$ and $x = 6$

$$\frac{\Delta R(t)}{\Delta t} = \frac{200(0.8909)^6 - 200(0.8909)^4}{2} = -12.996$$

Since rates of change are increasing, the graph of $R(t)$ is concave up.

**53.** **(a)** Since $f(x)$ is exponential, its formula will be $f(x) = ab^x$. Since $f(0) = 0.5$,

$$f(0) = ab^0 = 0.5.$$

But $b^0 = 1$, so

$$a(1) = 0.5$$
$$a = 0.5.$$

We now know that $f(x) = 0.5b^x$. Since $f(1) = 2$, we have

$$f(1) = 0.5b^1 = 2$$
$$0.5b = 2$$
$$b = 4$$

So $f(x) = 0.5(4)^x$.

We will find a formula for $g(x)$ the same way.

$$g(x) = ab^x.$$

Since $g(0) = 4$,

$$g(0) = ab^0 = 4$$
$$a = 4.$$

Therefore,

$$g(x) = 4b^x.$$

We'll use $g(2) = \frac{4}{9}$ to get

$$g(2) = 4b^2 = \frac{4}{9}$$
$$b^2 = \frac{1}{9}$$
$$b = \pm\frac{1}{3}.$$

Since $b > 0$,

$$g(x) = 4\left(\frac{1}{3}\right)^x.$$

Since $h(x)$ is linear, its formula will be

$$h(x) = b + mx.$$

We know that $b$ is the y-intercept, which is 2, according to the graph. Since the points $(a, a + 2)$ and $(0, 2)$ lie on the graph, we know that the slope, $m$, is

$$\frac{(a + 2) - 2}{a - 0} = \frac{a}{a} = 1,$$

so the formula is

$$h(x) = 2 + x.$$

**(b)** We begin with

$$f(x) = g(x)$$
$$\frac{1}{2}(4)^x = 4\left(\frac{1}{3}\right)^x.$$

Since the variable is an exponent, we need to use logs, so

$$\log\left(\frac{1}{2} \cdot 4^x\right) = \log\left(4 \cdot \left(\frac{1}{3}\right)^x\right)$$
$$\log\frac{1}{2} + \log(4)^x = \log 4 + \log\left(\frac{1}{3}\right)^x$$
$$\log\frac{1}{2} + x\log 4 = \log 4 + x\log\frac{1}{3}.$$

Now we will move all expressions containing the variable to one side of the equation:

$$x\log 4 - x\log\frac{1}{3} = \log 4 - \log\frac{1}{2}.$$

Factoring out $x$, we get

$$x\left(\log 4 - \log\frac{1}{3}\right) = \log 4 - \log\frac{1}{2}$$
$$x\log\left(\frac{4}{1/3}\right) = \log\left(\frac{4}{1/2}\right)$$
$$x\log 12 = \log 8$$
$$x = \frac{\log 8}{\log 12}.$$

This is the exact value of $x$. Note that $\frac{\log 8}{\log 12} \approx 0.837$, so $f(x) = g(x)$ when $x$ is exactly $\frac{\log 8}{\log 12}$ or about 0.837.

**(c)** Since $f(x) = h(x)$, we want to solve

$$\frac{1}{2}(4)^x = x + 2.$$

The variable does not occur only as an exponent, so logs cannot help us solve this equation. Instead, we need to graph the two functions and note where they intersect. The points occur when $x \approx 1.378$ or $x \approx -1.967$.

**57. (a)** Applying the given formula,

$$\text{Number toads in year 0 is } P = \frac{1000}{1 + 49(1/2)^0} = 20$$
$$\text{Number toads in year 5 is } P = \frac{1000}{1 + 49(1/2)^5} = 395$$
$$\text{Number toads in year 10 is } P = \frac{1000}{1 + 49(1/2)^{10}} = 954.$$

**(b)** We set up and solve the equation $P = 500$:

$$\begin{aligned}
\frac{1000}{1+49(1/2)^t} &= 500 \\
500\left(1+49(1/2)^t\right) &= 1000 && \text{multiplying by denominator} \\
1+49(1/2)^t &= 2 && \text{dividing by 500} \\
49(1/2)^t &= 1 \\
(1/2)^t &= 1/49 \\
\log(1/2)^t &= \log(1/49) && \text{taking logs} \\
t\log(1/2) &= \log(1/49) && \text{using a log rule} \\
t &= \frac{\log(1/49)}{\log(1/2)} \\
&= 5.615,
\end{aligned}$$

and so it takes about 5.6 years for the population to reach 500. A similar calculation shows that it takes about 7.2 years for the population to reach 750.

**(c)** The graph in Figure 5.1 suggests that the population levels off at about 1000 toads. We can see this algebraically by using the fact that $(1/2)^t \to 0$ as $t \to \infty$. Thus,

$$P = \frac{1000}{1+49(1/2)^t} \to \frac{1000}{1+0} = 1000 \text{ toads} \qquad \text{as} \quad t \to \infty.$$

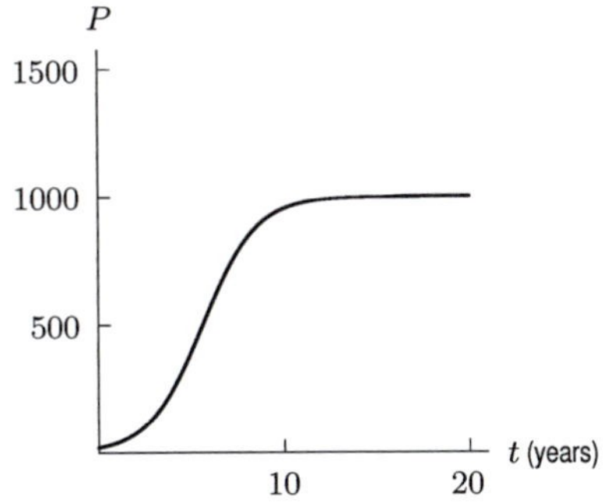


**Figure 5.1**: Toad population, $P$, against time, $t$

**61. (a)** We have

$$\begin{aligned}
1.12^t &= 6.3 \\
\left(10^{\log 1.12}\right)^t &= 10^{\log 6.3} \\
10^{t\cdot\overbrace{\log 1.12}^{v}} &= 10^{\overbrace{\log 6.3}^{w}},
\end{aligned}$$

so $v = \log 1.12$ and $w = \log 6.3$.

**(b)** We see that

$$\begin{aligned}
10^{vt} &= 10^{w} \\
vt &= w && \text{same base, so exponents are equal} \\
t &= \frac{w}{v}.
\end{aligned}$$

A numerical approximation is given by

$$t = \frac{\log 1.12}{\log 6.3}$$
$$= 16.241.$$

Checking out answer, we see that $1.12^{16.241} = 6.300$, as required.

## Solutions for Section 5.3

### Skill Refresher

**S1.** $\log 0.0001 = \log 10^{-4} = -4 \log 10 = -4$.

**S5.** The equation $-\ln x = 12$ can be expressed as $\ln x = -12$, which is equivalent to $x = e^{-12}$.

**S9.** Rewrite the sum as $\ln x^3 + \ln x^2 = \ln(x^3 \cdot x^2) = \ln x^5$.

### Exercises

**1.** The graphs of $y = 10^x$ and $y = 2^x$ both have horizontal asymptotes, $y = 0$. The graph of $y = \log x$ has a vertical asymptote, $x = 0$.

**5.** See Figure 5.2. The graph of $y = 2 \cdot 3^x + 1$ is the graph of $y = 3^x$ stretched vertically by a factor of 2 and shifted up by 1 unit.

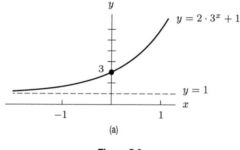

(a)

**Figure 5.2**

**9.** A graph of this function is shown in Figure 5.3. We see that the function has a vertical asymptote at $x = 3$. The domain is $(3, \infty)$.

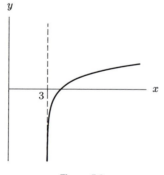

**Figure 5.3**

**13. (a)** Since $\log x$ becomes more and more negative as $x$ decreases to 0 from above,

$$\lim_{x \to 0^+} \log x = -\infty.$$

**(b)** Since $-x$ is positive if $x$ is negative and $-x$ decreases to 0 as $x$ increases to 0 from below,

$$\lim_{x \to 0^-} \ln(-x) = -\infty.$$

**17.** We know, by the definition of pH, that $4.5 = -\log[H^+]$. Therefore, $-4.5 = \log[H^+]$, and $10^{-4.5} = [H^+]$. Thus, the hydrogen ion concentration is $10^{-4.5} = 3.162 \times 10^{-5}$ moles per liter.

## Problems

**21. (a)** Let the functions graphed in (a), (b), and (c) be called $f(x), g(x)$, and $h(x)$ respectively. Looking at the graph of $f(x)$, we see that $f(10) = 3$. In the table for $r(x)$ we note that $r(10) = 1.699$ so $f(x) \neq r(x)$. Similarly, $s(10) = 0.699$, so $f(x) \neq s(x)$. The values describing $t(x)$ do seem to satisfy the graph of $f(x)$, however. In the graph, we note that when $0 < x < 1$, then $y$ must be negative. The data point $(0.1, -3)$ satisfies this. When $1 < x < 10$, then $0 < y < 3$. In the table for $t(x)$, we see that the point $(2, 0.903)$ satisfies this condition. Finally, when $x > 10$ we see that $y > 3$. The values $(100, 6)$ satisfy this. Therefore, $f(x)$ and $t(x)$ could represent the same function.

**(b)** For $g(x)$, we note that

$$\begin{cases} \text{when } 0 < x < 0.2, & \text{then } y < 0; \\ \text{when } 0.2 < x < 1, & \text{then } 0 < y < 0.699; \\ \text{when } x > 1, & \text{then } y > 0.699. \end{cases}$$

All the values of $x$ in the table for $r(x)$ are greater than 1 and all the corresponding values of $y$ are greater than 0.699, so $g(x)$ could equal $r(x)$. We see that, in $s(x)$, the values $(0.5, -0.060)$ do not satisfy the second condition so $g(x) \neq s(x)$. Since we already know that $t(x)$ corresponds to $f(x)$, we conclude that $g(x)$ and $r(x)$ correspond.

**(c)** By elimination, $h(x)$ must correspond to $s(x)$. We see that in $h(x)$,

$$\begin{cases} \text{when } x < 2, & \text{then } y < 0; \\ \text{when } 2 < x < 20, & \text{then } 0 < y < 1; \\ \text{when } x > 20, & \text{then } y > 1. \end{cases}$$

Since the values in $s(x)$ satisfy these conditions, it is reasonable to say that $h(x)$ and $s(x)$ correspond.

**25.** Let $I_A$ and $I_B$ be the intensities of sound $A$ and sound $B$, respectively. We know that $I_B = 5I_A$ and, since sound $A$ measures 30 dB, we know that $10 \log(I_A/I_0) = 30$. We have:

$$\begin{aligned} \text{Decibel rating of B} &= 10 \log\left(\frac{I_B}{I_0}\right) \\ &= 10 \log\left(\frac{5I_A}{I_0}\right) \\ &= 10 \log 5 + 10 \log\left(\frac{I_A}{I_0}\right) \\ &= 10 \log 5 + 30 \\ &= 10(0.699) + 30 \\ &\approx 37. \end{aligned}$$

Notice that although sound B is 5 times as loud as sound A, the decibel rating only goes from 30 to 37.

**29.** We know $M_1 = \log\left(\frac{W_1}{W_0}\right)$ and $M_2 = \log\left(\frac{W_2}{W_0}\right)$. Thus,

$$\begin{aligned} M_2 - M_1 &= \log\left(\frac{W_2}{W_0}\right) - \log\left(\frac{W_1}{W_0}\right) \\ &= \log\left(\frac{W_2}{W_1}\right). \end{aligned}$$

33. (a) The pH is 2.3, which, according to our formula for pH, means that

$$-\log\left[H^+\right] = 2.3.$$

This means that

$$\log\left[H^+\right] = -2.3.$$

This tells us that the exponent of 10 that gives $[H^+]$ is $-2.3$, so

$$\left[H^+\right] = 10^{-2.3} \qquad \text{because } -2.3 \text{ is exponent of } 10$$
$$= 0.005 \text{ moles/liter.}$$

(b) From part (a) we know that 1 liter of lemon juice contains 0.005 moles of $H^+$ ions. To find out how many $H^+$ ions our lemon juice has, we need to convert ounces of juice to liters of juice and moles of ions to numbers of ions. We have

$$2 \text{ oz} \times \frac{1 \text{ liter}}{30.3 \text{ oz}} = 0.066 \text{ liters.}$$

We see that

$$0.066 \text{ liters juice} \times \frac{0.005 \text{ moles } H^+ \text{ ions}}{1 \text{ liter}} = 3.3 \times 10^{-4} \text{ moles } H^+ \text{ ions.}$$

There are $6.02 \times 10^{23}$ ions in one mole, and so

$$3.3 \times 10^{-4} \text{ moles} \times \frac{6.02 \times 10^{23} \text{ ions}}{\text{mole}} = 1.987 \times 10^{20} \text{ ions.}$$

37. A possible formula is $y = \ln x$.

## Solutions for Section 5.4

### Skill Refresher

S1. $1.455 \times 10^6$

S5. $3.6 \times 10^{-4}$

S9. Since $0.1 < \frac{1}{3} < 1$, $10^{-1} < \frac{1}{3} < 1 = 10^0$.

### Exercises

1. Using a linear scale, the wealth of everyone with less than a million dollars would be indistinguishable because all of them are less than one one-thousandth of the wealth of the average billionaire. A log scale is more useful.

5. (a)

Table 5.1

| $n$ | 1 | 2 | 3 | 4 | 5 | 6 | 7 | 8 | 9 |
|-----|---|---|---|---|---|---|---|---|---|
| $\log n$ | 0 | 0.3010 | 0.4771 | 0.6021 | 0.6990 | 0.7782 | 0.8451 | 0.9031 | 0.9542 |

Table 5.2

| $n$ | 10 | 20 | 30 | 40 | 50 | 60 | 70 | 80 | 90 |
|-----|----|----|----|----|----|----|----|----|----|
| $\log n$ | 1 | 1.3010 | 1.4771 | 1.6021 | 1.6990 | 1.7782 | 1.8451 | 1.9031 | 1.9542 |

(b) The first tick mark is at $10^0 = 1$. The dot for the number 2 is placed $\log 2 = 0.3010$ of the distance from 1 to 10. The number 3 is placed at $\log 3 = 0.4771$ units from 1, and so on. The number 30 is placed 1.4771 units from 1, the number 50 is placed 1.6989 units from 1, and so on.

**Figure 5.4**

## Problems

**9.** The Declaration of Independence was signed in 1776, about 225 years ago. We can write this number as

$$\frac{225}{1,000,000} = 0.000225 \text{ million years ago.}$$

This number is between $10^{-4} = 0.0001$ and $10^{-3} = 0.001$. Using a calculator, we have

$$\log 0.000225 \approx -3.65,$$

which, as expected, lies between $-3$ and $-4$ on the log scale. Thus, the Declaration of Independence is placed at

$$10^{-3.65} \approx 0.000224 \text{ million years ago} = 224 \text{ years ago.}$$

**13. (a)** Figure 5.5 shows the track events plotted on a linear scale.

**Figure 5.5:**

**(b)** Figure 5.6 shows the track events plotted on a logarithmic scale.

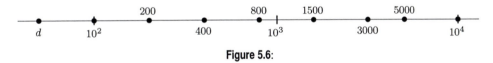

**Figure 5.6:**

**(c)** Figure 5.5 gives a runner better information about pacing for the distance.
**(d)** On Figure 5.5 the point 50 is $\frac{1}{2}$ the distance from 0 to 100. On Figure 5.6 the point 50 is the same distance to the left of 100 as 200 is to the right. This is shown as point $d$.

**17. (a)**

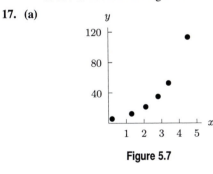

**Figure 5.7**

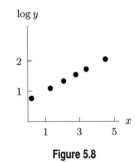

**Figure 5.8**

(b) The data appear to be exponential.

(c) See Figure 5.8. The data appear to be linear.

**Table 5.3**

| $x$ | 0.2 | 1.3 | 2.1 | 2.8 | 3.4 | 4.5 |
|---|---|---|---|---|---|---|
| $\log y$ | .76 | 1.09 | 1.33 | 1.54 | 1.72 | 2.05 |

**21.** (a) Find the values of $\ln t$ in the table, use linear regression on a calculator or computer with $x = \ln t$ and $y = P$. The line has slope $-7.787$ and $P$-intercept $86.283$ ($P = -7.787 \ln t + 86.283$). Thus $a = -7.787$ and $b = 86.283$.

(b) Figure 5.9 shows the data points plotted with $P$ against $\ln t$. The model seems to fit well.

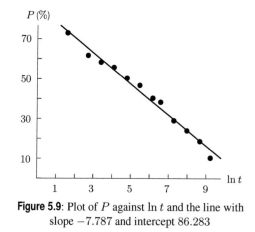

**Figure 5.9**: Plot of $P$ against $\ln t$ and the line with slope $-7.787$ and intercept $86.283$

(c) The subjects will recognize no words when $P = 0$, that is, when $-7.787 \ln t + 86.283 = 0$. Solving for $t$:

$$-7.787 \ln t = -86.283$$
$$\ln t = \frac{86.283}{7.787}$$

Taking both sides to the $e$ power,

$$e^{\ln t} = e^{\frac{86.283}{7.787}}$$
$$t \approx 64{,}918.342,$$

so $t \approx 45$ days.

The subject recognized all the words when $P = 100$, that is, when $-7.787 \ln t + 86.283 = 100$. Solving for $t$:

$$-7.787 \ln t = 13.717$$
$$\ln t = \frac{13.717}{-7.787}$$
$$t \approx 0.172,$$

so $t \approx 0.172$ minutes ($\approx 10$ seconds) from the start of the experiment.

**(d)**

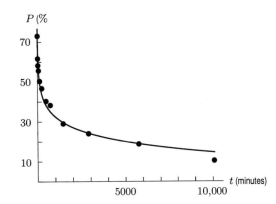

**Figure 5.10**: The percentage $P$ of words recognized as
a function of $t$, the time elapsed and the function
$P = -7.787 \ln t + 86.283$

# Solutions for Chapter 5 Review

## Exercises

**1.** The continuous percent growth rate is the value of $k$ in the equation $Q = ae^{kt}$, which is $-10$.

To convert to the form $Q = ab^t$, we first say that the right sides of the two equations equal each other (since each equals $Q$), and then we solve for $a$ and $b$. Thus, we have $ab^t = 7e^{-10t}$. At $t = 0$, we can solve for $a$:

$$ab^0 = 7e^{-10\cdot 0}$$
$$a \cdot 1 = 7 \cdot 1$$
$$a = 7.$$

Thus, we have $7b^t = 7e^{-10t}$, and we solve for $b$:

$$7b^t = e^{-10t}$$
$$b^t = e^{-10t}$$
$$b^t = \left(e^{-10}\right)^t$$
$$b = e^{-10} \approx 0.0000454.$$

Therefore, the equation is $Q = 7(0.0000454)^t$.

**5.** To convert to the form $Q = ae^{kt}$, we first say that the right sides of the two equations equal each other (since each equals $Q$), and then we solve for $a$ and $k$. Thus, we have $ae^{kt} = 4 \cdot 8^{1.3t}$. At $t = 0$, we can solve for $a$:

$$ae^{k\cdot 0} = 4 \cdot 8^0$$
$$a \cdot 1 = 4 \cdot 1$$
$$a = 4.$$

Thus, we have $4e^{kt} = 4 \cdot 8^{1.3t}$, and we solve for $k$:

$$4e^{kt} = 4 \cdot 8^{1.3t}$$
$$e^{kt} = \left(8^{1.3}\right)^t$$

$$\left(e^k\right)^t = 14.929^t$$
$$e^k = 14.929$$
$$\ln e^k = \ln 14.929$$
$$k = 2.703.$$

Therefore, the equation is $Q = 4e^{2.703t}$.

**9.** Since the goal is to get $t$ by itself as much as possible, first divide both sides by 3, and then use logs.

$$3(1.081)^t = 14$$
$$1.081^t = \frac{14}{3}$$
$$\log(1.081)^t = \log(\frac{14}{3})$$
$$t \log 1.081 = \log(\frac{14}{3})$$
$$t = \frac{\log(\frac{14}{3})}{\log 1.081}.$$

**13.**

$$5(1.031)^x = 8$$
$$1.031^x = \frac{8}{5}$$
$$\log(1.031)^x = \log \frac{8}{5}$$
$$x \log 1.031 = \log \frac{8}{5} = \log 1.6$$
$$x = \frac{\log 1.6}{\log 1.031} = 15.395.$$

Check your answer: $5(1.031)^{15.395} \, approx \, 8$.

**17.**

$$3^{(4 \log x)} = 5$$
$$\log 3^{(4 \log x)} = \log 5$$
$$(4 \log x) \log 3 = \log 5$$
$$4 \log x = \frac{\log 5}{\log 3}$$
$$\log x = \frac{\log 5}{4 \log 3}$$
$$x = 10^{\frac{\log 5}{4 \log 3}} \approx 2.324$$

**21.**

$$\frac{\log x^2 + \log x^3}{\log(100x)} = 3$$
$$\log x^2 + \log x^3 = 3 \log(100x)$$
$$2 \log x + 3 \log x = 3(\log 100 + \log x)$$
$$5 \log x = 3(2 + \log x)$$
$$5 \log x = 6 + 3 \log x$$

$$2 \log x = 6$$
$$\log x = 3$$
$$x = 10^3 = 1000.$$

To check, we see that

$$\frac{\log x^2 + \log x^3}{\log(100x)} = \frac{\log(1000^2) + \log(1000^3)}{\log(100 \cdot 1000)}$$
$$= \frac{\log(1,000,000) + \log(1,000,000,000)}{\log(100,000)}$$
$$= \frac{6+9}{5}$$
$$= 3,$$

as required.

**25.** Using the fact that $A^{-1} = 1/A$ and the log rules:

$$\ln(A+B) - \ln(A^{-1} + B^{-1}) = \ln(A+B) - \ln\left(\frac{1}{A} + \frac{1}{B}\right)$$
$$= \ln(A+B) - \ln\frac{A+B}{AB}$$
$$= \ln\left((A+B) \cdot \frac{AB}{A+B}\right)$$
$$= \ln(AB).$$

**29.**  • The natural logarithm is defined only for positive inputs, so the domain of this function is given by

$$300 - x > 0$$
$$x < 300.$$

 • The graph of $y = \ln x$ has a vertical asymptote at $x = 0$, that is, where the input is zero. So the graph of $y = \ln(300 - x)$ has a vertical asymptote where its input is zero, at $x = 300$.

**33.** We have $\log 4838 = 3.685$, so this lifespan would be marked at 3.7 inches.

**37.** We have $\log 4 = 0.602$, so this lifespan would be marked at 0.6 inches.

## Problems

**41. (a)**

$$e^{x+3} = 8$$
$$\ln e^{x+3} = \ln 8$$
$$x + 3 = \ln 8$$
$$x = \ln 8 - 3 \approx -0.9206$$

**(b)**

$$4(1.12^x) = 5$$
$$1.12^x = \frac{5}{4} = 1.25$$
$$\log 1.12^x = \log 1.25$$
$$x \log 1.12 = \log 1.25$$
$$x = \frac{\log 1.25}{\log 1.12} \approx 1.9690$$

(c)

$$e^{-0.13x} = 4$$
$$\ln e^{-0.13x} = \ln 4$$
$$-0.13x = \ln 4$$
$$x = \frac{\ln 4}{-0.13} \approx -10.6638$$

(d)

$$\log(x - 5) = 2$$
$$x - 5 = 10^2$$
$$x = 10^2 + 5 = 105$$

(e)

$$2\ln(3x) + 5 = 8$$
$$2\ln(3x) = 3$$
$$\ln(3x) = \frac{3}{2}$$
$$3x = e^{\frac{3}{2}}$$
$$x = \frac{e^{\frac{3}{2}}}{3} \approx 1.4939$$

(f)

$$\ln x - \ln(x - 1) = \frac{1}{2}$$
$$\ln\left(\frac{x}{x-1}\right) = \frac{1}{2}$$
$$\frac{x}{x-1} = e^{\frac{1}{2}}$$
$$x = (x - 1)e^{\frac{1}{2}}$$
$$x = xe^{\frac{1}{2}} - e^{\frac{1}{2}}$$
$$e^{\frac{1}{2}} = xe^{\frac{1}{2}} - x$$
$$e^{\frac{1}{2}} = x(e^{\frac{1}{2}} - 1)$$
$$\frac{e^{\frac{1}{2}}}{e^{\frac{1}{2}} - 1} = x$$
$$x \approx 2.5415$$

Note: (g) (h) and (i) can not be solved analytically, so we use graphs to approximate the solutions.

(g) From Figure 5.11 we can see that $y = e^x$ and $y = 3x + 5$ intersect at $(2.534, 12.601)$ and $(-1.599, 0.202)$, so the values of $x$ which satisfy $e^x = 3x + 5$ are $x = 2.534$ or $x = -1.599$. We also see that $y_1 \approx 12.601$ and $y_2 \approx 0.202$.

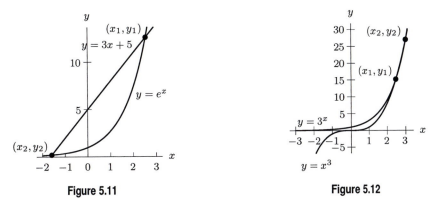

Figure 5.11                                      Figure 5.12

(h) The graphs of $y = 3^x$ and $y = x^3$ are seen in Figure 5.12. It is very hard to see the points of intersection, though $(3, 27)$ would be an immediately obvious choice (substitute 3 for $x$ in each of the formulas). Using technology, we can find a second point of intersection, $(2.478, 15.216)$. So the solutions for $3^x = x^3$ are $x = 3$ or $x = 2.478$.

Since the points of intersection are very close, it is difficult to see these intersections even by zooming in. So, alternatively, we can find where $y = 3^x - x^3$ crosses the $x$-axis. See Figure 5.13.

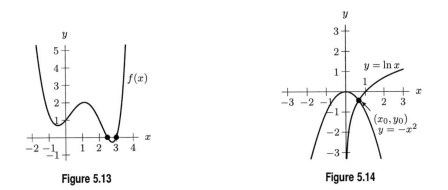

Figure 5.13                                      Figure 5.14

(i) From the graph in Figure 5.14, we see that $y = \ln x$ and $y = -x^2$ intersect at $(0.6529, -0.4263)$, so $x = 0.6529$ is the solution to $\ln x = -x^2$.

**45.** We have

$$M_2 - M_1 = \log\left(\frac{W_2}{W_1}\right)$$

$$5.6 - 4.4 = \log\left(\frac{W_2}{W_1}\right)$$

$$1.2 = \log\left(\frac{W_2}{W_1}\right)$$

$$\frac{W_2}{W_1} = 10^{1.2} = 15.849$$

$$W_2 = 15.849 \cdot W_1.$$

The seismic waves of the second earthquake are about 15.85 times larger.

**49. (a)** The number of bacteria present after 1/2 hour is

$$N = 1000e^{0.69(1/2)} \approx 1412.$$

If you notice that $0.69 \approx \ln 2$, you could also say

$$N = 1000e^{0.69/2} \approx 1000e^{\frac{1}{2}\ln 2} = 1000e^{\ln 2^{1/2}} = 1000e^{\ln \sqrt{2}} = 1000\sqrt{2} \approx 1412.$$

**(b)** We solve for $t$ in the equation

$$1,000,000 = 1000e^{0.69t}$$
$$e^{0.69t} = 1000$$
$$0.69t = \ln 1000$$
$$t = \left(\frac{\ln 1000}{0.69}\right) \approx 10.011 \text{ hours.}$$

**(c)** The doubling time is the time $t$ such that $N = 2000$, so

$$2000 = 1000e^{0.69t}$$
$$e^{0.69t} = 2$$
$$0.69t = \ln 2$$
$$t = \left(\frac{\ln 2}{0.69}\right) \approx 1.005 \text{ hours.}$$

If you notice that $0.69 \approx \ln 2$, you see why the half-life turns out to be 1 hour:

$$e^{0.69t} = 2$$
$$e^{t \ln 2} \approx 2$$
$$e^{\ln 2^t} \approx 2$$
$$2^t \approx 2$$
$$t \approx 1$$

**53.** We have

$$v(t) = 30$$
$$20e^{0.2t} = 30$$
$$e^{0.2t} = \frac{30}{20}$$
$$0.2t = \ln\left(\frac{30}{20}\right)$$
$$t = \frac{\ln\left(\frac{30}{20}\right)}{0.2} = 2.027.$$

**57. (a)** For $f(x) = 10^x$,

Domain of $f(x)$ is all $x$

Range of $f(x)$ is all $y > 0$.

There is one asymptote, the horizontal line $y = 0$.

**(b)** Since $g(x) = \log x$ is the inverse function of $f(x) = 10^x$, the domain of $g(x)$ corresponds to range of $f(x)$ and range of $g(x)$ corresponds to domain of $g(x)$.

Domain of $g(x)$ is all $x > 0$

Range of $g(x)$ is all $y$.

The asymptote of $f(x)$ becomes the asymptote of $g(x)$ under reflection across the line $y = x$. Thus, $g(x)$ has one asymptote, the line $x = 0$.

**61. (a)** Since the drug is being metabolized continuously, the formula for describing the amount left in the bloodstream is $Q(t) = Q_0e^{kt}$. We know that we start with 2 mg, so $Q_0 = 2$, and the rate of decay is 4%, so $k = -0.04$. (Why is $k$ negative?) Thus $Q(t) = 2e^{-0.04t}$.

**(b)** To find the percent decrease in one hour, we need to rewrite our equation in the form $Q = Q_0 b^t$, where $b$ gives us the percent left after one hour:

$$Q(t) = 2e^{-0.04t} = 2(e^{-0.04})^t \approx 2(0.96079)^t.$$

We see that $b \approx 0.96079 = 96.079\%$, which is the percent we have left after one hour. Thus, the drug level decreases by about $3.921\%$ each hour.

**(c)** We want to find out when the drug level reaches $0.25$ mg. We therefore ask when $Q(t)$ will equal $0.25$.

$$2e^{-0.04t} = 0.25$$
$$e^{-0.04t} = 0.125$$
$$-0.04t = \ln 0.125$$
$$t = \frac{\ln 0.125}{-0.04} \approx 51.986.$$

Thus, the second injection is required after about $52$ hours.

**(d)** After the second injection, the drug level is $2.25$ mg, which means that $Q_0$, the initial amount, is now $2.25$. The decrease is still $4\%$ per hour, so when will the level reach $0.25$ again? We need to solve the equation

$$2.25e^{-0.04t} = 0.25,$$

where $t$ is now the number of hours since the second injection.

$$e^{-0.04t} = \frac{0.25}{2.25} = \frac{1}{9}$$
$$-0.04t = \ln(1/9)$$
$$t = \frac{\ln(1/9)}{-0.04} \approx 54.931.$$

Thus the third injection is required about $55$ hours after the second injection, or about $52 + 55 = 107$ hours after the first injection.

**65. (a)** We have

$$\sqrt{\log(\text{googol})} = \sqrt{\log(10^{100})}$$
$$= \sqrt{100}$$
$$= 10.$$

**(b)** We have

$$\log\sqrt{\text{googol}} = \log\sqrt{10^{100}}$$
$$= \log\left((10^{100})^{0.5}\right)$$
$$= \log(10^{50})$$
$$= 50.$$

**(c)** We have

$$\sqrt{\log(\text{googolplex})} = \sqrt{\log\left(10^{\text{googol}}\right)}$$
$$= \sqrt{\text{googol}}$$
$$= \sqrt{10^{100}}$$
$$= (10^{100})^{0.5}$$
$$= 10^{50}.$$

# CHECK YOUR UNDERSTANDING

1. False. Since the $\log 1000 = \log 10^3 = 3$ we know $\log 2000 > 3$. Or use a calculator to find that $\log 2000$ is about 3.3.

5. True. Comparing the equation, we see $b = e^k$, so $k = \ln b$.

9. True. The log function outputs the power of 10 which in this case is $n$.

13. False. For example, $\log 10 = 1$, but $\ln 10 \approx 2.3$.

17. True. The two functions are inverses of one another.

21. False. Taking the natural log of both sides we see $t = \ln 7.32$ .

25. True. This is the definition of half-life.

29. True. Solve for $t$ by dividing both sides by $Q_0$, taking the ln of both sides and then dividing by $k$.

33. False. Since $26,395,630,000,000 \approx 2.6 \cdot 10^{13}$, we see that it would be between 13 and 14 on a log scale.

37. False. The fit will not be as good as $y = x^3$ but an exponential function can be found.

## Solutions to Skills for Chapter 5

1. $\log(\log 10) = \log(1) = 0$.

5. $\dfrac{\log 1}{\log 10^5} = \dfrac{0}{5 \log 10} = \dfrac{0}{5} = 0$.

9. The equation $10^{-4} = 0.0001$ is equivalent to $\log 0.0001 = -4$.

13. The equation $\log 0.01 = -2$ is equivalent to $10^{-2} = 0.01$.

17. The expression is not the logarithm of a quotient, so it cannot be rewritten using the properties of logarithms.

21. There is no rule for the logarithm of a sum, it cannot be rewritten.

25. Rewrite the sum as $\log 12 + \log x = \log 12x$.

29. Rewrite with powers and combine,

$$
\begin{aligned}
3\left(\log(x+1) + \frac{2}{3}\log(x+4)\right) &= 3\log(x+1) + 2\log(x+4) \\
&= \log(x+1)^3 + \log(x+4)^2 \\
&= \log\left((x+1)^3(x+4)^2\right)
\end{aligned}
$$

33. The logarithm of a sum cannot be simplified.

37. Rewrite as $\log 100^{2z} = 2z \log 100 = 2z(2) = 4z$.

41. We divide both sides by 3 to get

$$5^x = 3.$$

Taking logs of both sides we get

$$\log(5^x) = \log 3.$$

This gives

$$x \log 5 = \log 3$$

or in other words

$$x = \frac{\log 3}{\log 5} \approx 0.683.$$

**45.** Taking logs of both sides we get

$$\log 19^{6x} = \log(77 \cdot 7^{4x}).$$

This gives

$$6x \log 19 = \log 77 + \log 7^{4x}$$
$$6x \log 19 = \log 77 + 4x \log 7$$
$$6x \log 19 - 4x \log 7 = \log 77$$
$$x(6 \log 19 - 4 \log 7) = \log 77$$
$$x = \frac{\log 77}{6 \log 19 - 4 \log 7} \approx 0.440.$$

**49.** We first re-arrange the equation so that the natural log is alone on one side, and we then convert to exponential form:

$$2 \ln(6x - 1) + 5 = 7$$
$$2 \ln(6x - 1) = 2$$
$$\ln(6x - 1) = 1$$
$$e^{\ln(6x-1)} = e^1$$
$$6x - 1 = e$$
$$6x = e + 1$$
$$x = \frac{e + 1}{6} \approx 0.620.$$

# CHAPTER SIX

## Solutions for Section 6.1

### Skill Refresher

**S1.** Substituting $x = 4$ into $f(x)$, we have $f(4) = \sqrt{4} = 2$.

**S5.** We know $e^x = 1$ has only one solution, $x = 0$.

**S9.** (a) The translation $y = f(x - 4) = \ln(x - 4)$ shifts the graph to the right 4 units.
   (b) The translation $y = f(x) - 7 = \ln(x) - 7$ shifts the graph down 7 units.
   (c) The translation $y = f(x + \sqrt{2}) = \ln(x + \sqrt{2})$ shifts the graph to the left $\sqrt{2}$ units.
   (d) The translation $y = f(x - 3) + 5 = \ln(x - 3) + 5$ shifts the graph to the right 3 units, and shifts the graph up 5 units.

### Exercises

**1. (a)**

| $x$ | $-1$ | 0 | 1 | 2 | 3 |
|---|---|---|---|---|---|
| $g(x)$ | $-3$ | 0 | 2 | 1 | $-1$ |

The graph of $g(x)$ is shifted one unit to the right of $f(x)$.

**(b)**

| $x$ | $-3$ | $-2$ | $-1$ | 0 | 1 |
|---|---|---|---|---|---|
| $h(x)$ | $-3$ | 0 | 2 | 1 | $-1$ |

The graph of $h(x)$ is shifted one unit to the left of $f(x)$.

**(c)**

| $x$ | $-2$ | $-1$ | 0 | 1 | 2 |
|---|---|---|---|---|---|
| $k(x)$ | 0 | 3 | 5 | 4 | 2 |

The graph $k(x)$ is shifted up three units from $f(x)$.

**(d)**

| $x$ | $-1$ | 0 | 1 | 2 | 3 |
|---|---|---|---|---|---|
| $m(x)$ | 0 | 3 | 5 | 4 | 2 |

The graph $m(x)$ is shifted one unit to the right and three units up from $f(x)$.

**5.** See Figure 6.1.

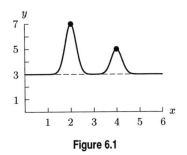

**Figure 6.1**

**9.** The graph of $g(x)$ is shifted four units to the left of $f(x)$, and the graph of $h(x)$ is shifted two units to the right of $f(x)$.

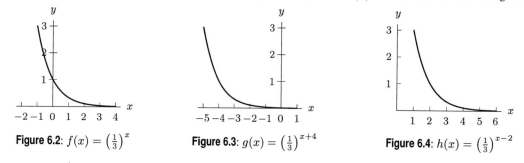

**Figure 6.2:** $f(x) = \left(\frac{1}{3}\right)^x$     **Figure 6.3:** $g(x) = \left(\frac{1}{3}\right)^{x+4}$     **Figure 6.4:** $h(x) = \left(\frac{1}{3}\right)^{x-2}$

**13.** $m(n) + 1 = \dfrac{1}{2}n^2 + 1$

To graph this function, shift the graph of $m(n) = \frac{1}{2}n^2$ one unit up. See Figure 6.5.

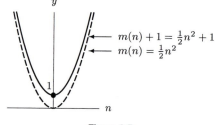

**Figure 6.5**

**17.** $m(n) + \sqrt{13} = \dfrac{1}{2}n^2 + \sqrt{13}$

To sketch, shift the graph of $m(n) = \frac{1}{2}n^2$ up by $\sqrt{13}$ units, as in Figure 6.6.

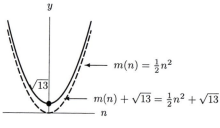

**Figure 6.6**

**21.** $k(w) - 3 = 3^w - 3$

To sketch, shift the graph of $k(w) = 3^w$ down 3 units, as in Figure 6.7.

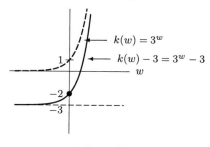

**Figure 6.7**

**25.** $k(w + 2.1) - 1.3 = 3^{w+2.1} - 1.3$

To sketch, shift the graph of $k(w) = 3^w$ to the left by 2.1 units and down 1.3 units, as in Figure 6.8.

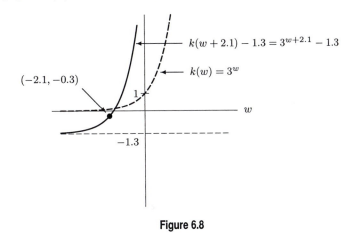

**Figure 6.8**

## Problems

**29.** The graph in Figure 6.10 is a result of shifting the function in Figure 6.9 right 2 units and down 6 units. Thus $y = f(x - 2) - 6$ is a formula for the function in Figure 6.10.

**33. (a)** $P(t) + 100$ describes a population that is always 100 people larger than the original population.

**(b)** $P(t + 100)$ describes a population that has the same number of people as the original population, but the number occurs 100 years earlier.

**37.** To compensate for the down shift, we shift up 1. To compensate for the left shift by 3, we shift right by 3.

**41.** The graph of $y = x^2 - 10x + 25$ appears to be the graph of $y = x^2$ moved to the right by 5 units. See Figure 6.9. If this were so, then its formula would be $y = (x - 5)^2$. Since $(x - 5)^2 = x^2 - 10x + 25$, $y = x^2 - 10x + 25$ is, indeed, a horizontal shift of $y = x^2$.

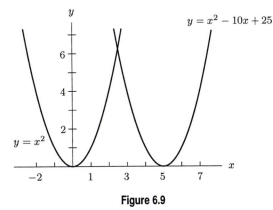

**Figure 6.9**

**45.** Since this is an inside change, the graph is four units to the left of $q(z)$. That is, for any given $z$ value, the value of $q(z+4)$ is the same as the value of the function $q$ evaluated four units to the right of $z$ (at $z + 4$).

**49. (a)** On day $d$, high tide in Tacoma, $T(d)$, is 1 foot higher than high tide in Seattle, $S(d)$. Thus, $T(d) = S(d) + 1$.

**(b)** On day $d$, height of the high tide in Portland equals high tide of the previous day, i.e. $d - 1$, in Seattle. Thus, $P(d) = S(d - 1)$.

**53. (a)** If each drink costs \$7 then $x$ drinks cost \$7$x$. Adding this to the \$20 cover charge gives $20 + 7x$. So

$$t(x) = 20 + 7x, \quad \text{for } x \geq 0.$$

**(b)** The cover charge is now \$25, so we have

$$n(x) = 25 + 7x$$
$$= 5 + \underbrace{20 + 7x}$$
$$= 5 + t(x).$$

Alternatively, notice that for any number of drinks the new cost, $n(x)$, is \$5 more than the old cost, $t(x)$. So

$$n(x) = t(x) + 5.$$

Thus $n(x)$ is the vertical shift of $t(x)$ up 5 units.

**(c)** Since 2 drinks are free, a customer who orders $x$ drinks pays for only $(x - 2)$ drinks at \$7/drink if $x \geq 2$. Thus

$$p(x) = 30 + 7(x - 2), \quad \text{if } x \geq 2.$$

The formula for $p(x)$ if $x \geq 2$ can be written in terms of $t(x)$ as follows:

$$p(x) = 10 + \underbrace{20 + 7(x - 2)}_{t(x-2)}$$
$$= 10 + t(x - 2) \text{ if } x \geq 2.$$

Another way to think of this is to subtract two from your total number of drinks, $x$. Use $t(x-2)$ to determine the cost of two fewer drinks with the initial cover charge. Then add this 10 dollar increase in the cover charge to the result, so $p(x) = t(x - 2) + 10$. This shows that the cover charge is \$10 more but you are charged for 2 fewer drinks.

**57.** Subtract the normal temperature from the thermometer reading. The fever is

$$f(t) = H(t) - 37 \text{ degrees}.$$

## Solutions for Section 6.2

### Skill Refresher

**S1.** Evaluating $f(x)$ at $x = 3$, we have $f(3) = e^3 = 20.086$.

**S5. (a)** $f(-x) = 2(-x)^2 = 2x^2$
  **(b)** $-f(x) = -\left(2x^2\right) = -2x^2$
  Notice since $f(-x) = f(x) = 2x^2$, we see that $f(x)$ is an even function.

**S9. (a)** $f(-x) = 3(-x)^4 - 2(-x) = 3x^4 + 2x$
  **(b)** $-f(x) = -\left(3x^4 - 2x\right) = -3x^4 + 2x$
  Notice since $f(-x) \neq -f(x)$ and $f(-x) \neq -f(x)$, we see that $f(x) = 3x^4 - 2x$ is neither an even nor an odd function.

### Exercises

**1. (a)** The $y$-coordinate is unchanged, but the $x$-coordinate is the same distance to the left of the $y$-axis, so the point is $(-2, -3)$.
  **(b)** The $x$-coordinate is unchanged, but the $y$-coordinate is the same distance above the $x$-axis, so the point is $(2, 3)$.

**5.** The negative sign reflects the graph of $Q(t)$ horizontally about the $y$-axis, so the domain of $y = Q(-t)$ is $t < 0$. A horizontal reflection of the graph of $Q(t)$ about the $y$-axis will not change the range. The range of $Q(-t)$ is therefore the same as the range of $Q(t)$, $-4 \leq Q(-t) \leq 7$.

**9.** To reflect about the $x$-axis, we make all the $y$-values negative, getting $y = -e^x$ as the formula.

**13.** The graph of $y = -g(x) = -(1/3)^x$ is the graph of $y = g(x)$ reflected across the $x$-axis. See Figure 6.10.

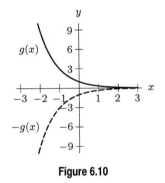

**Figure 6.10**

**17.** We have

$$y = m(-n) + 3 = (-n)^2 - 4(-n) + 5 + 3$$
$$= n^2 + 4n + 8.$$

To graph this function, first reflect the graph of $m$ across the $y$-axis, and then shift it up by 3 units. See Figure 6.11.

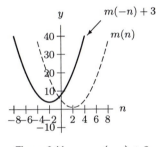

**Figure 6.11**: $y = m(-n) + 3$

**21.** We have

$$y = -k(w - 2) = -3^{w-2}$$

To graph this function, first reflect the graph of $k$ across the $w$-axis, then shift it to the right by 2 units. See Figure 6.12.

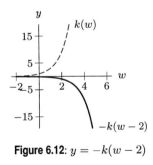

**Figure 6.12**: $y = -k(w - 2)$

**25.** Since $f(-x) = (-x)^5 + 3(-x)^3 - 2 = -x^5 - 3x^3 - 2$ is equal to neither $f(x)$ or $-f(x)$, the function is neither even nor odd.

## Problems

**29. (a)** $g(-x) = \sqrt[3]{-x} = -\sqrt[3]{x} = -g(x)$.
 **(b)** From part (a), we see that $g(-x) = -g(x)$, so the graphs of these two functions coincide in Figure 6.13.

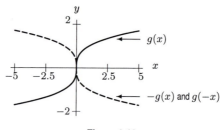

**Figure 6.13**

 **(c)** The function is odd, since $g(-x) = -g(x)$.

**33.** The answers are

(i)  b                    (ii)  c                    (iii)  d
(iv)  e                   (v)  a

**37.** Since $f(x)$ is an even function we know $f(-x) = f(x)$. Similarly since $g(x)$ is an odd function we know $g(-x) = -g(x)$. Using both of these identities, we evaluate at $-x$.
 **(a)**

$$h(-x) = f(-x)g(-x)$$
$$= f(x)(-g(x))$$
$$= -f(x)g(x)$$
$$= -h(x).$$

Since $h(-x) = -h(x)$, the function $h(x) = f(x)g(x)$ must be odd.
 **(b)**

$$k(-x) = f(-x) + g(-x)$$
$$= f(x) + (-g(x))$$
$$= f(x) - g(x).$$

Since $k(-x) = f(x) - g(x)$ and $k(x) = f(x) + g(x)$, it follows $k(-x) = k(x)$ only if $g(x) = 0$, and $k(-x) = -k(x)$ only if $f(x) = 0$. Therefore unless $f(x) = 0$ or $g(x) = 0$, $k(x)$ will be neither even nor odd.
 **(c)**

$$m(-x) = g(f(-x))$$
$$= g(f(x))$$
$$= m(x).$$

Since $m(-x) = m(x)$, the function $m(x) = g(f(x))$ must be even.

**41.** The argument that $f(x)$ is not odd is correct. However, the statement "something is either even or odd" is false. This function is neither an odd function nor an even function.

**45.** Because $f(x)$ is an odd function, $f(x) = -f(-x)$. Setting $x = 0$ gives $f(0) = -f(0)$, so $f(0) = 0$. Since $c(0) = 1$, $c(x)$ is not odd. Since $d(0) = 1$, $d(x)$ is not odd.

**49.** Suppose $f(x)$ is both even and odd. If $f(x)$ is even, then

$$f(-x) = f(x).$$

If $f(x)$ is odd, then

$$f(-x) = -f(x).$$

Since $f(-x)$ equals both $f(x)$ and $-f(x)$, we have

$$f(x) = -f(x).$$

Add $f(x)$ to both sides of the equation to get

$$2f(x) = 0$$

or

$$f(x) = 0.$$

Thus, the function $f(x) = 0$ is the only function which is both even and odd. There are no *nontrivial* functions that have both symmetries.

## Solutions for Section 6.3

### Skill Refresher

**S1.** (a) In order to evaluate $2f(6)$, we first evaluate $f(6) = 6^2 = 36$. Then we multiply by 2, and thus we have $2f(6) = 2 \cdot 36 = 72$.

(b) Since $f(6) = 36$, we have $-\frac{1}{2}f(6) = -\frac{1}{2} \cdot 36 = -18$.

(c) Since $f(6) = 36$, we have $5f(6) - 3 = 5(36) - 3 = 177$.

(d) In order to evaluate $\frac{1}{4}f(x-1)$ at $x = 6$, we first evaluate $f(6-1) = f(5) = 5^2 = 25$. Next we divide by 4, and thus we have $\frac{1}{4}f(6-1) = \frac{1}{4} \cdot 25 = \frac{25}{4}$.

### Exercises

**1.** To increase by a factor of 10, multiply by 10. The right shift of 2 is made by substituting $x - 2$ for $x$ in the function formula. Together they give $y = 10f(x - 2)$.

**5.** Since the domain of $R(n)$ is the same as the domain of $P(n)$, no horizontal transformations have been applied. Since the maximum value of $R(n)$ is $-5$ times the minimum value of $P(n)$, and the minimum value $R(n)$ approaches is $-5$ times the maximum value $P(n)$ approaches, $P(n)$ has been stretched vertically by a factor of 5 and reflected about the $x$-axis. Thus, we have

$$R(n) = -5P(n)$$

.

**9.** See Figure 6.14.

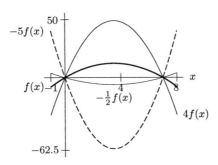

**Figure 6.14**

**13. (a)**

**Table 6.1**

| $x$ | $-4$ | $-3$ | $-2$ | $-1$ | 0 | 1 | 2 | 3 | 4 |
|---|---|---|---|---|---|---|---|---|---|
| $f(-x)$ | 13 | 6 | 1 | $-2$ | $-3$ | $-2$ | 1 | 6 | 13 |

**(b)**

**Table 6.2**

| $x$ | $-4$ | $-3$ | $-2$ | $-1$ | 0 | 1 | 2 | 3 | 4 |
|---|---|---|---|---|---|---|---|---|---|
| $-f(x)$ | $-13$ | $-6$ | $-1$ | 2 | 3 | 2 | $-1$ | $-6$ | $-13$ |

**(c)**

**Table 6.3**

| $x$ | $-4$ | $-3$ | $-2$ | $-1$ | 0 | 1 | 2 | 3 | 4 |
|---|---|---|---|---|---|---|---|---|---|
| $3f(x)$ | 39 | 18 | 3 | $-6$ | $-9$ | $-6$ | 3 | 18 | 39 |

**(d)** All three functions are even.

**17.** As before, $g(x) = x^2$. Thus $y = g(-x) = (-x)^2$, but $(-x)^2 = x^2$, so $g(-x) = x^2 = g(x)$. Since $g(x)$ is an even function, reflecting its graph across the $y$-axis leaves the graph unchanged. See Figure 6.15.

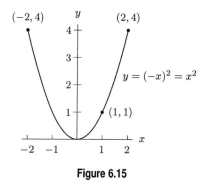

**Figure 6.15**

## Problems

**21.** The factor of 2 doubles the resulting $y$-values (stretching the graph vertically). Then the input $(x + 1)$ moves the graph 1 unit to the left.

**25.** The graph of $0.5f(t)$ resembles a flattened version of the graph of $f(t)$. Each point on the new graph is half as far from the $t$-axis as the same point on the graph of $f$. See Figure 6.16.

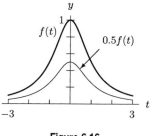

**Figure 6.16**

**29.** I is (b)

    II is (d)

    III is (c)

    IV is (h)

**33.** See Figure 6.17. The graph is shifted to the right by 3 units.

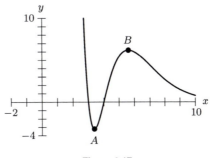

**Figure 6.17**

**37.** **(a)** Notice that the value of $h(x)$ at every value of $x$ is one-half the value of $f(x)$ at the same $x$ value. Thus, $f(x)$ has been compressed vertically by a factor of $1/2$, and

$$h(x) = \frac{1}{2}f(x).$$

    **(b)** Observe that $k(-6) = f(6)$, $k(-4) = f(4)$, and so on. Thus, we have

$$k(x) = f(-x).$$

    **(c)** The values of $m(x)$ are 4 less than the values of $f(x)$ at the same $x$ value. Thus, we have

$$m(x) = f(x) - 4.$$

**41.** Figure 6.18 gives a graph of a function $y = f(x)$ together with graphs of $y = \frac{1}{2}f(x)$ and $y = 2f(x)$. All three graphs cross the $x$-axis at $x = -2, x = -1$, and $x = 1$. Likewise, all three functions are increasing and decreasing on the same intervals. Specifically, all three functions are increasing for $x < -1.55$ and for $x > 0.21$ and decreasing for $-1.55 < x < 0.21$.

    Even though the stretched and compressed versions of $f$ shown by Figure 6.18 are increasing and decreasing on the same intervals, they are doing so at different rates. You can see this by noticing that, on every interval of $x$, the graph of $y = \frac{1}{2}f(x)$ is less steep than the graph of $y = f(x)$. Similarly, the graph of $y = 2f(x)$ is steeper than the graph of $y = f(x)$. This indicates that the magnitude of the average rate of change of $y = \frac{1}{2}f(x)$ is less than that of $y = f(x)$, and that the magnitude of the average rate of change of $y = 2f(x)$ is greater than that of $y = f(x)$.

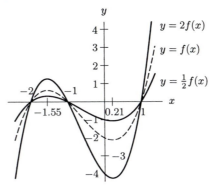

**Figure 6.18**: The graph of $y = 2f(x)$ and $y = \frac{1}{2}f(x)$ compared to the graph of $f(x)$

## Solutions for Section 6.4

### Skill Refresher

**S1.** We have $f(2x) = (2x)^3 - 5 = 8x^3 - 5$.

**S5.** $Q\left(\frac{1}{3}t\right) = 4e^{6(\frac{1}{3}t)} = 4e^{2t}$.

### Exercises

1. The graph is compressed horizontally by a factor of $1/2$ so the transformed function gives an output of 3 when the input is 1. Thus the point $(1, 3)$ lies on the graph of $g(2x)$.

5. The graph in Figure 6.19 of $n(x) = e^{2x}$ is a horizontal compression of the graph of $m(x) = e^x$. The graph of $p(x) = 2e^x$ is a vertical stretch of the graph of $m(x) = e^x$. All three graphs have a horizontal asymptote at $y = 0$. The $y$-intercept of $n(x) = e^{2x}$ is the same as for $m(x)$, but the graph of $p(x) = 2e^x$ has a $y$-intercept of $(0, 2)$.

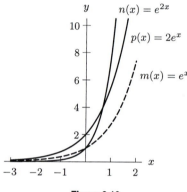

**Figure 6.19**

9. See Figure 6.20.

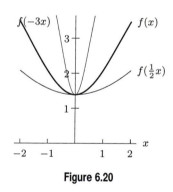

**Figure 6.20**

### Problems

13. **(a)** The graph is compressed horizontally by a factor of $\frac{1}{2}$, so the new domain is $-6 \leq x \leq 6$. There is no outside change so the range is still $0 \leq l(2x) \leq 3$.

    **(b)** The graph is stretched horizontally by a factor of 2, so the new domain is $-24 \leq x \leq 24$. The change is only an inside change, so the range is still $0 \leq l(\frac{1}{2}x) \leq 3$.

**17.** Since $x$ km is $1000x$ meters, the temperature at a depth of $x$ km is

$$p(x) = T(1000x).$$

**21.** If profits are $r(t) = 0.5P(t)$ instead of $P(t)$, then profits are half the dollar level expected. If profits are $s(t) = P(0.5t)$ instead of $P(t)$, then profits are accruing half as fast as the projected rate.

**25.** The function $f$ has been reflected over the $x$-axis and the $y$-axis and stretched horizontally by a factor of 2. Thus, $y = -f(-\frac{1}{2}x)$.

**29. (a)** Since $f(x)$ has been stretched horizontally by a factor of 4, the domain of $h(x)$ is $-24 \le x \le 8$.

**(b)** Since the average rate of change of $f(x)$ over $-6 \le x \le 2$ is given as $-3$, we have

$$\frac{f(2) - f(-6)}{2 - (-6)} = \text{ Average rate of change of } f(x)$$

$$\frac{f(2) - f(-6)}{8} = -3$$

$$f(2) - f(-6) = 8(-3) = -24.$$

We now use the fact that $f(2) - f(-6) = -24$ to calculate the average rate of change of $h(x)$ over its domain $-24 \le x \le 8$.

$$\begin{aligned}
\text{Average rate of change of } h(x) &= \frac{\Delta y}{\Delta x} = \frac{h(8) - h(-24)}{8 - (-24)} \\
&= \frac{f\left(\frac{1}{4}(8)\right) - f\left(\frac{1}{4}(-24)\right)}{32} \quad \text{(since } h(x) = f(1/4 \cdot x)\text{)} \\
&= \frac{f(2) - f(-6)}{32} \\
&= \frac{-24}{32} \quad \text{(since } f(2) - f(-6) = -24\text{)} \\
&= -\frac{3}{4}.
\end{aligned}$$

Notice the average rate of change of $h(x) = f\left(\frac{1}{4}x\right)$ over the stretched interval $-24 \le x \le 8$ is one-fourth of the average rate of change of $f(x)$ over the original interval $-6 \le x \le 2$ since $\Delta y$ has not changed while $\Delta x$ has been stretched by a factor of 4.

# Solutions for Section 6.5

## Skill Refresher

**S1.** Factoring 4 out from the left side of the equation, we have

$$4(x + 3) = 4(x - h).$$

Thus, we see $h = -3$.

**S5. (a)**

$$f(8(1)) - 3 = f(8) - 3 = \sqrt[3]{8} - 3 = 2 - 3 = -1$$

**(b)**

$$8f(1 - 3) = 8f(-2) = 8\sqrt[3]{-2}$$

**(c)**

$$8f(1) - 3 = 8\sqrt[3]{1} - 3 = 8(1) - 3 = 5$$

**(d)**

$$8\left(f(1) - 3\right) = 8(\sqrt[3]{1} - 3) = 8(-2) = -16$$

**(e)**

$$f(8(1 - 3)) = f(8(-2)) = f(-16) = \sqrt[3]{-16} = -2\sqrt[3]{2}$$

**(f)**

$$f(8(1) - 3) = f(5) = \sqrt[3]{5}$$

**S9.** We must write the expression $-1/3x - 9$ inside the function in the form $B(x - h)$. Factoring out the $-1/3$, we have $-1/3x - 9 = -1/3(x + 27)$. Thus, we have

$$y = 6f\left(-\frac{1}{3}x - 9\right) = 6f\left(-\frac{1}{3}(x + 27)\right).$$

We see from the form above that $A = 6$, $B = -1/3$, $h = -27$, and $k = 0$.

## Exercises

1. We transform the function $f(x)$ horizontally since all of the operations occur inside the function $f(x)$. To identify the transformations, we first factor out 3 in the expression $3x - 2$:

$$y = f(3x - 2) = f(3(x - \frac{2}{3})).$$

So there are two inside transformations applied to $f$: First a horizontal compression by a factor of $1/3$ and then a horizontal shift right by $2/3$ units.

5. We first write the function $w(t) = 40 - 2v(-0.5t)$ in the form $w(t) = Av(b(t - h)) + k$, and we have

$$w(t) = 40 - 2v(-0.5t) = -2v(-0.5t) + 40.$$

We have vertically stretched $v(t)$ by a factor of 2 and reflected about the $t$-axis, then horizontally stretched by a factor of 2 and reflected about the $v$-axis, and finally vertically shifted up by 40 units.

Notice a result of the horizontal transformations (stretch and reflection), the new domain of $w(t)$ is $t = -8, -6, -4, -2,$ and 0. Evaluating $w(t)$ at these values, we have

$$\begin{aligned}
w(-8) &= -2v\left(-0.5(-8)\right) + 40 \\
&= -2v(4) + 40 \\
&= -2(23) + 40 \\
&= -6,
\end{aligned}$$

$$\begin{aligned}
w(-6) &= -2v\left(-0.5(-6)\right) + 40 \\
&= -2v(3) + 40 \\
&= -2(19) + 40 \\
&= 2,
\end{aligned}$$

and so. Table 6.4 gives the rest of the values of $w(t)$.

**Table 6.4**

| $t$ | $-8$ | $-6$ | $-4$ | $-2$ | 0 |
|---|---|---|---|---|---|
| $w(t)$ | $-6$ | 2 | 8 | 6 | 0 |

**9.** The graph is stretched vertically by a factor of 2, then stretched horizontally by a factor of 2, an finally shifted vertically by 20 units. See Figure 6.21.

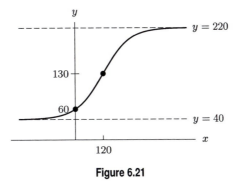

**Figure 6.21**

**13.** We have $w(t) = 3 - 0.5v(-2t) = -0.5v(-2t) + 3$.

The inside changes only affect the graph's $x$-coordinates, and so they do not affect its horizontal asymptote, since horizontally shifting or stretching a horizontal line has no effect.

- First, compress the graph horizontally by a factor of $1/2$. This moves the vertical asymptote from $t = 5$ to $t = 2.5$.
- Next, flip the graph horizontally across the $y$-axis. This moves the vertical asymptote from $t = 2.5$ to $t = -2.5$.

The outside changes only affect the graph's $y$-coordinates, and so they do not affect its vertical asymptote, since vertically shifting or stretching a vertical line has no effect.

- First, compress vertically by a factor of $1/2$. This moves the horizontal asymptote from $y = -4$ to $y = -2$.
- Next, flip the graph vertically across the $t$-axis. This moves the horizontal asymptote from $y = -2$ to $y = 2$.
- Finally, shift the graph vertically by 3 units. This moves the horizontal asymptote from $y = 2$ to $y = 5$.

Thus, the graph of $w$ has vertical asymptote $t = -2.5$ and horizontal asymptote $y = 5$.

## Problems

**17. (a)**

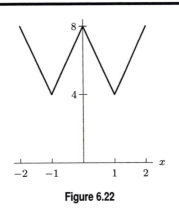

**Figure 6.22**

**(b)**

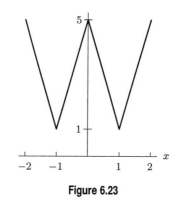

Figure 6.23

**(c)** Switching the order of transformations in parts (a) and (b) leads to two different graphs. Notice in (a), Figure 6.22 has maximum value $y = 8$ and minimum value $y = 4$. In (b), Figure 6.23 has maximum value $y = 5$ and minimum value $y = 1$. In both cases however, the difference between the maximum and minimum values is 4.

**21.** We have

$$y = f(x + 4) \qquad \text{shift left 4 units}$$
$$y = -f(x + 4) \qquad \text{then flip vertically}$$
$$y = -f(x + 4) + 2 \qquad \text{then shift up 2 units}$$
$$y = 3\left(-f(x + 4) + 2\right) \quad \text{then stretch by 3}$$

so    $g(x) = -3f(x + 4) + 6.$

**25.** We see that

$$y = h(x + 3) \qquad \text{shift left 3 units}$$
$$y = 2h(x + 3) \qquad \text{stretch vertically by 2}$$
$$y = 2h(x + 3) + 6 \quad \text{shift vertically by 6}$$

so    $f(x) = 2h(x + 3) + 6.$

Since $(-12, 20)$ is on the graph of $f$, this means

$$f(-12) = 2h(-12 + 3) + 6 = 20$$
$$2h(-9) + 6 \qquad = 20$$
$$2h(-9) \qquad = 14$$
$$h(-9) \qquad = 7.$$

Thus, $(-9, 7)$ is a point on the graph of $h$. Likewise, since $(0, 6)$ and $(36, -2)$ are points on the graph of $f$, we have

$$f(0) = 2h(0 + 3) + 6 \ = 6$$
$$2h(3) + 6 \qquad = 6$$
$$2h(3) \qquad = 0$$
$$h(3) \qquad = 0 \qquad \text{so } (3, 0) \text{ is on graph of } h$$
$$f(36) = 2h(36 + 3) + 6 = -2$$
$$2h(39) + 6 \qquad = -2$$
$$2h(39) \qquad = -8$$
$$h(39) \qquad = -4. \quad \text{so } (39, -4) \text{ is on graph of } h$$

Thus, points on the graph of $h$ include $(-9, 7), (3, 0), (39, -4)$. To check our answer, if $(-9, 7)$ is on the graph of $h$ and we shift it to the left 3 units, then $(-12, 7)$ is the new point. If we stretch this vertically by a factor of 2, we get the point $(-12, 14)$. Finally, if we shift it up by 6, we get the point $(-12, 20)$, which is in fact on the graph of $f$, as required. We can perform a similar check with the other two points.

**29.** All four transformations are equivalent.

**(a)** We have:

$$
\begin{array}{lll}
\text{First step:} & y = f(x) + 3 & \text{shift up} \\
\text{Second step:} & y = 2\left(f(x) + 3\right) & \text{stretch} \\
\text{Third step:} & y = -2\left(f(x) + 3\right). & \text{flip}
\end{array}
$$

Multiplying out, this gives $y = -2f(x) - 6$.

**(b)** We have:

$$
\begin{array}{lll}
\text{First step:} & y = -f(x) & \text{flip} \\
\text{Second step:} & y = -f(x) - 3 & \text{shift down} \\
\text{Third step:} & y = 2(-f(x) - 3). & \text{stretch}
\end{array}
$$

Multiplying out, this gives $y = -2f(x) - 6$, which is the same as part (a).

**(c)** We have:

$$
\begin{array}{lll}
\text{First step:} & y = 2f(x) & \text{stretch} \\
\text{Second step:} & y = 2f(x) + 6 & \text{shift up} \\
\text{Third step:} & y = -\left(2f(x) + 6\right). & \text{flip}
\end{array}
$$

Multiplying out, this gives $y = -2f(x) - 6$, which is the same as parts (a) and (b).

**(d)** We have:

$$
\begin{array}{lll}
\text{First step:} & y = -f(x) & \text{flip} \\
\text{Second step:} & y = -2f(x) & \text{stretch} \\
\text{Third step:} & y = -2f(x) - 6. & \text{shift down}
\end{array}
$$

This is the same as parts (a)–(c).

**33.** A horizontal shift of $g$ to the right $k$ units gives

$$
\begin{aligned}
y &= g(x - k) \\
&= e^{(x-k)} \\
&= e^x \cdot e^{-k} \quad \text{(since } e^{ab} = e^a \cdot e^b\text{)} \\
&= e^{-k} \cdot g(x).
\end{aligned}
$$

If $k > 0$, then $0 < e^{-k} < 1$. Thus we see shifting $g$ right $k$ units is equivalent to vertically compressing of $g$ by a factor of $e^{-k}$.

**37. (a)** The building is kept at $60°$ F until 5 am when the heat is turned up. The building heats up at a constant rate until 7 am when it is $68°$ F. It stays at that temperature until 3 pm when the heat is turned down. The building cools at a constant rate until 5 pm. At that time, the temperature is $60°$ F and it stays that level through the end of the day.

**(b)** Since $c(t) = 142 - d(t) = -d(t) + 142$, the graph of $c(t)$ will look like the graph of $d(t)$ that has been first vertically reflected across the $t$-axis and then vertically shifted up 142 units.

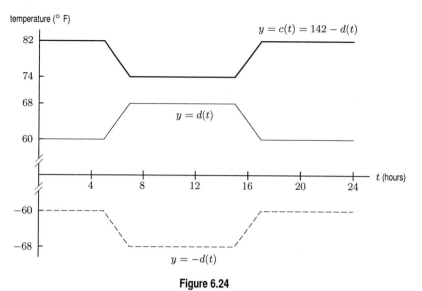

**Figure 6.24**

(c) This could describe the cooling schedule in the summer months when the temperature is kept at $82°$ F at night and cooled down to $74°$ during the day.

**41.** (a) Vertically stretching the graph of $f(x) = x$ by a factor of $m > 1$ generates the family of functions $y = mx$, that is, straight lines with slope $m > 1$ through the origin. Since $m$ is greater than 1, this family is all lines through the origin with slope greater than 1.

(b) Vertical stretching and horizontally shifting the graph of $f(x) = x$ generates the family of functions $y = m(x-h) = mx - mh$. If we write $-mh = b$, then we have $y = mx + b$, that is, the family of straight lines with slope $m > 1$ and $y$-intercept $b$. Since $m$ is greater than 1 and $b$ can have any value, this is the entire family of increasing linear functions with slope greater than 1. The family in part (a) is contained in the family found in part (b).

**45.** Yes, $g$ is also quadratic. We see that:

$$
\begin{aligned}
f(sx) &= a(sx - h)^2 + k & \text{because } f(x) = a(x-h)^2 + k \\
&= a\left(s(x - h/s)\right)^2 + k & \text{factor out } s \\
&= as^2(x - h/s)^2 + k
\end{aligned}
$$

which means

$$
\begin{aligned}
g(x) &= r \cdot f(sx) + j \\
&= r\left(as^2(x - h/s)^2 + k\right) + j \\
&= \underbrace{ras^2}_{A} \cdot \left(x - \underbrace{h/s}_{H}\right)^2 + \underbrace{rk + j}_{K}
\end{aligned}
$$

so

$$
\begin{aligned}
A &= ras^2 \\
H &= h/s \\
K &= rk + j.
\end{aligned}
$$

# Solutions for Chapter 6 Review

## Exercises

**1.** (a) The input is $2x = 2 \cdot 2 = 4$.
   (b) The input is $\frac{1}{2}x = \frac{1}{2} \cdot 2 = 1$.
   (c) The input is $x + 3 = 2 + 3 = 5$.
   (d) The input is $-x = -2$.

**5.** A function is odd if $a(-x) = -a(x)$.

$$a(x) = \frac{1}{x}$$
$$a(-x) = \frac{1}{-x} = -\frac{1}{x}$$
$$-a(x) = -\frac{1}{x}$$

Since $a(-x) = -a(x)$, we know that $a(x)$ is an odd function.

**9.** A function is even if $b(-x) = b(x)$.

$$b(x) = |x|$$
$$b(-x) = |-x| = |x|$$

Since $b(-x) = b(x)$, we know that $b(x)$ is an even function.

## Problems

**13.** The graph is the graph of $f$ shifted to the left by 2 and up by 2. See Figure 6.25.

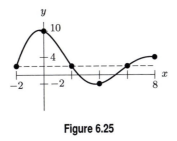

**Figure 6.25**

**17. (a)** Since the $x$-coordinate of the point $(-3, 1)$ on the graph of $f(x)$ has been multiplied by $-1$ in order to obtain the point $(3, 1)$ on the graph of $g(x)$, $g(x)$ must be obtained by reflecting the graph of $f(x)$ horizontally about the $y$-axis.

**(b)** Since the $y$-coordinate remains constant and only the $x$-coordinate is moved, $f(x)$ must be shifted horizontally. The $x$-coordinate of the point $(3, 1)$ on the graph of $g(x)$ has been shifted to the right by 6 units from the point $(-3, 1)$ on the graph of $f(x)$.

**21. (a)** Using the formula for $d(t)$, we have

$$d(t) - 15 = (-16t^2 + 38) - 15$$
$$= -16t^2 + 23.$$

$$d(t - 1.5) = -16(t - 1.5)^2 + 38$$
$$= -16(t^2 - 3t + 2.25) + 38$$
$$= -16t^2 + 48t + 2.$$

**(b)**

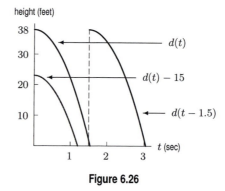

height (feet)

$d(t)$

$d(t) - 15$

$d(t - 1.5)$

$t$ (sec)

**Figure 6.26**

**(c)** $d(t) - 15$ represents the height of a brick which falls from $38 - 15 = 23$ feet above the ground. On the other hand, $d(t - 1.5)$ represents the height of a brick which began to fall from 38 feet above the ground at one and a half seconds after noon.

**(d)**   (i) The brick hits the ground when its height is 0. Thus, if we represent the brick's height above the ground by $d(t)$, we get

$$0 = d(t)$$
$$0 = -16t^2 + 38$$
$$-38 = -16t^2$$
$$t^2 = \frac{38}{16}$$
$$t^2 = 2.375$$
$$t = \pm\sqrt{2.375} \approx \pm 1.541.$$

We are only interested in positive values of $t$, so the brick must hit the ground 1.541 seconds after noon.

(ii) If we represent the brick's height above the ground by $d(t) - 15$ we get

$$0 = d(t) - 15$$
$$0 = -16t^2 + 23$$
$$-23 = -16t^2$$
$$t^2 = \frac{23}{16}$$
$$t^2 = 1.4375$$
$$t = \pm\sqrt{1.4375} \approx \pm 1.199.$$

Again, we are only interested in positive values of $t$, so the the brick hits the ground 1.199 seconds after noon.

**(e)** Since the brick, whose height is $d(t - 1.5)$, begins falling 1.5 seconds after the brick whose height is $d(t)$, we expect the brick whose height is $d(t - 1.5)$ to hit the ground 1.5 seconds after the brick whose height is $d(t)$. Thus, the brick should hit the ground $1.5 + 1.541 = 3.041$ seconds after noon.

**25.** The graph appears to have been shifted to the left 6 units, compressed vertically by a factor of 2, and shifted vertically by 1 unit, so

$$y = \frac{1}{2}h(x + 6) + 1.$$

**29.**

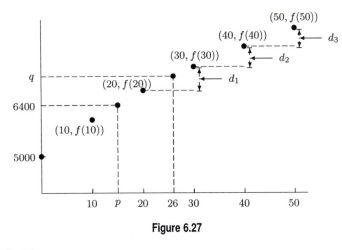

**Figure 6.27**

**33.** There is a vertical stretch of 3 so

$$y = 3h(x).$$

**37.** 
- The vertical gap between the horizontal asymptote and the lowest point (at the $y$-intercept) is 1 in the original graph but 2 in the new graph. Thus, the graph appears to have been vertically stretched by a factor of 2.
- The two points where the graph crosses the horizontal asymptote are separated by a distance of 2 in the old graph but by a distance of 4 in the new graph. So, the graph appears to have been horizontally stretched by a factor of 2.
- The horizontal asymptote is at $y = 0$ in the old graph but $y = 3$ in the new graph, so the graph appears to have been vertically shifted by 3 units.

  Putting all this together gives $y = 2f(x/2) + 3$.

**41.** Temperatures in this borehole are the same as temperatures 5 meters deeper in the Belleterre borehole. See Table 6.5.

**Table 6.5**

| $d$ | 20 | 45 | 70 | 95 | 120 | 145 | 170 | 195 |
|---|---|---|---|---|---|---|---|---|
| $h(d)$ | 5.5 | 5.2 | 5.1 | 5.1 | 5.3 | 5.5 | 5.75 | 6 |

**45.** Temperatures in this borehole are 2°C warmer than temperatures 50% higher than temperatures at the same depth in the Belleterre borehole. See Table 6.6.

**Table 6.6**

| $d$ | 25 | 50 | 75 | 100 | 125 | 150 | 175 | 200 |
|---|---|---|---|---|---|---|---|---|
| $q(d)$ | 10.25 | 9.8 | 9.65 | 9.65 | 9.95 | 10.25 | 10.63 | 11 |

## CHECK YOUR UNDERSTANDING

**1.** True. The graph of $g(x)$ is a copy of the graph of $f$ shifted vertically up by three units.

**5.** True. The reflection across the $x$-axis of $y = f(x)$ is $y = -f(x)$.

**9.** True. Any point $(x, y)$ on the graph of $y = f(x)$ reflects across the $y$-axis to the point $(-x, y)$, which lies on the graph of $y = f(-x)$.

**13.** False. If $f(x) = x^2$, then $f(x + 1) = x^2 + 2x + 1 \neq x^2 + 1 = f(x) + 1$.

**17.** True.

**21.** False. Consider $f(x) = x^2$. Shifting up first and then compressing vertically gives the graph of $g(x) = \frac{1}{2}(x^2 + 1) = \frac{1}{2}x^2 + \frac{1}{2}$. Compressing first and then shifting gives the graph of $h(x) = \frac{1}{2}x^2 + 1$.

# CHAPTER SEVEN

## Solutions for Section 7.1

### Exercises

1. Graphs (I), (II), and (IV) appear to decribe period functions.

    (I) This function appears periodic. The rapid variation overlays a slower variation that appears to repeat every 8 units. (It almost appears to repeat every 4 units, but there is subtle difference between consecutive 4-second intervals. Do you see it?)

    (II) This function also appears period, again with a period of about 4 units. For instance, the $x$-intercepts appear to be evenly spaced, at approximately $-11, -7, -3, 1, 5, 9$, and the peaks are also evenly spaced, at $-9, -5, -1, 3, 7, 11$.

    (III) This function does not appear periodic. For instance, the $x$-intercepts grow increasingly close together (when read from left to right).

    (IV) At first glance this function might appear to vary unpredictably. But on closer inspection we see that the graph repeats the same pattern on the interval $-12 \le x \le 0$ and $0 \le x \le 12$.

    (V) This function does not appear periodic. The peaks of the graph appear to rise slowly (when read from left to right), and the troughs appear to fall slowly.

    (VI) This function does not appear to be periodic. The peaks and troughs of its graph seem to vary unpredictably, although they are more or less evenly spaced.

5. In the 9 o'clock position, the person is midway between the top and bottom of the wheel. Since the diameter is 150 m, the radius is 75 m, so the person is 75 m below the top, or $165 - 75 = 90$ m above the ground.

9. The period appears to be $41 - 1 = 40$.

### Problems

13. After 37 minutes, the person has completed one rotation. This means she is in the 6 o'clock position, at the bottom of the wheel. The diameter is 150 m, so the person is 150 m below the top, or $165 - 150 = 15$ m above the ground.

17. See Figure 7.1.

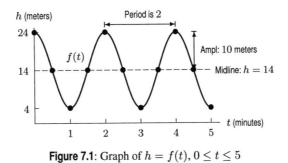

**Figure 7.1**: Graph of $h = f(t), 0 \le t \le 5$

21. Your initial position is twelve o'clock, since at $t = 0$, the value of $h$ is at its maximum of 35. The period is 4 because the wheel completes one cycle in 4 minutes. The diameter is 30 meters and the boarding platform is 5 meters above ground. Because you go through 2.5 cycles, the length of time spent on the wheel is 10 minutes.

**25.** The amplitude, period, and midline are the same for Figures 7.10 and 7.11 in the text. In Figure 7.11, the weight is initially moving upward toward the ceiling, since $d$, the distance from the ceiling, begins to decrease at $t = 0$, whereas in Figure 7.10, $d$ begins to increase at $t = 0$. Thus, the motion described in Figure 7.11 must have resulted from pulling the weight away from the ceiling at $t = -0.25$, whereas the motion described by Figure 7.10 must have resulted from pushing the weight toward the ceiling at $t = -0.25$.

**29.** Notice that the function is only approximately periodic. See Figure 7.2.

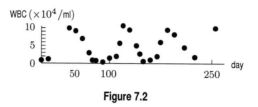

**Figure 7.2**

The midline is half way between the maximum and minimum WBC values.

$$y = \frac{(10.7 + 0.4)}{2} = 5.55.$$

The amplitude is the difference between the maximum and midline, so $A = 5.15$. The period is the length of time from peak to peak. Measuring between successive peaks gives $p_1 = 120 - 40 = 80$ days; $p_2 = 185 - 120 = 65$ days; $p_3 = 255 - 185 = 70$ days. Using the average of the three periods we get $p \approx 72$ days.

## Solutions for Section 7.2

### Exercises

**1.**

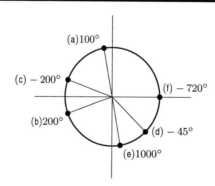

**Figure 7.3**

   **(a)** $(\cos 100°, \sin 100°) = (-0.174, 0.985)$
   **(b)** $(\cos 200°, \sin 200°) = (-0.940, -0.342)$
   **(c)** $(\cos(-200°), \sin(-200°)) = (-0.940, 0.342)$
   **(d)** $(\cos(-45°), \sin(-45°)) = (0.707, -0.707)$
   **(e)** $(\cos 1000°, \sin 1000°) = (0.174, -0.985)$
   **(f)** $(\cos 720°, \sin 720°) = (1, 0)$

**5.** To locate the points $D$, $E$, and $F$, we mark off their respective angles, $-90°$, $-135°$, and $-225°$, by measuring these angles from the positive $x$-axis in the clockwise direction. See Figure 7.4.

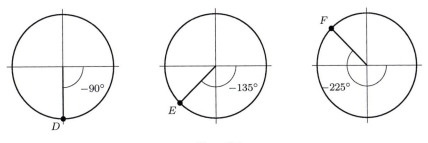

**Figure 7.4**

$$D = (0, -1),\ E = (-0.707, -0.707),\ F = (-0.707, 0.707)$$

**9.** We use the Pythagorean theorem to find the length of the hypotenuse:

$$\text{Hypotenuse}^2 = (0.1)^2 + (0.2)^2 = 0.01 + 0.04 = 0.05$$
$$\text{Hypotenuse} = \sqrt{0.05}.$$

**(a)** We have

$$\sin\theta = \frac{\text{Side opposite}}{\text{Hypotenuse}} = \frac{0.1}{\sqrt{0.05}} = 0.447.$$

**(b)** We have

$$\cos\theta = \frac{\text{Side adjacent}}{\text{Hypotenuse}} = \frac{0.2}{\sqrt{0.05}} = 0.894.$$

## Problems

**13.** See Figure 7.5.

  **(a)** $\cos 53°$ is positive, so we need an angle in the fourth quadrant with the same $x$-coordinate. This angle is $360° - 53° = 307°$.

  **(b)** $\sin 53°$ is positive, so we need an angle in the second quadrant with the same $y$-coordinate. This angle is $180° - 53° = 127°$.

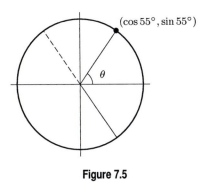

$(\cos 55°, \sin 55°)$

$\theta$

**Figure 7.5**

**17.** Given the angle $\theta$, draw a line $l$ through the origin making an angle $\theta$ with the $x$-axis. Go counterclockwise if $\theta > 0$ and clockwise if $\theta < 0$, wrapping around the unit circle more than once if necessary. Let $P = (x, y)$ be the point where $l$ intercepts the unit circle. Then the definition of sine is that $\sin\theta = y$.

**21. (a)** All sides have length 1, since triangle $\triangle KOL$ is an equilateral triangle. This is because all three angles are $60°$.

**(b)** Since triangles $\triangle OPK$ and $\triangle OPL$ are congruent, the length from $K$ to $P$ must be half the length of $KL$. Thus the length of $KP$ is $\frac{1}{2}(1) = 1/2$.

**(c)** Using the Pythagorean theorem we find that

$$\text{Distance from } O \text{ to } P = \sqrt{(\text{Length of hypotenuse})^2 - (\text{Distance from } K \text{ to } P)^2}$$
$$= \sqrt{1^2 - (1/2)^2}$$
$$= \sqrt{\frac{3}{4}}$$
$$= \frac{\sqrt{3}}{2}.$$

**(d)** Since $OP = \sqrt{3}/2$ and $KP = 1/2$, the coordinates of $K$ are $(\sqrt{3}/2, 1/2)$.

**(e)** It follows from part (d) that the cosine of $30°$ is $\sqrt{3}/2$ while the sine of $30°$ is $1/2$.

**(f)** In triangle $KOP$, we have $KP = 1/2$ and $OP = \sqrt{3}/2$. So

$$\sin 60° = \frac{\text{Opposite}}{\text{Hypotenuse}} = \frac{\sqrt{3}/2}{1} = \frac{\sqrt{3}}{2}$$
$$\cos 60° = \frac{\text{Adjacent}}{\text{Hypotenuse}} = \frac{1/2}{1} = \frac{1}{2}.$$

## Solutions for Section 7.3

### Exercises

**1.** Since the maximum value of the function is 3 and the minimum is 1, its midline is $y = 2$, and its amplitude is 1.

**5.** Since the middle of the clock's face is at 223 cm, the midline is at $i(t) = 223$ cm, and since the hand is 20 cm long, the amplitude is 20 cm.

**9.** Since the $x$-coordinate is $r \cos \theta$ and the $y$-coordinate is $r \sin \theta$ and $r = 3.8$ and $\theta = -180°$, the point is $(3.8 \cos(-180), 3.8 \sin(-180)) = (-3.8, 0)$.

**13.** Since the $x$-coordinate is $r \cos \theta$ and the $y$-coordinate is $r \sin \theta$ and $r = 3.8$ and $\theta = 1426°$, the point is $(3.8 \cos 1426, 3.8 \sin 1426) = (3.687, -0.919)$.

**17.** Since the $x$-coordinate is $r \cos \theta$ and the $y$-coordinate is $r \sin \theta$ and $r = 3.8$, the point is $(3.8 \cos 225, 3.8 \sin 225) = (-3.8\sqrt{2}/2, -3.8\sqrt{2}/2) = (-2.687, -2.687)$.

**21.**

$$x = r \cos \theta = 10 \cos 210° = 10(-\sqrt{3}/2) = -5\sqrt{3}$$

and

$$y = r \sin \theta = 10 \sin 210° = 10(-1/2) = -5,$$

so the coordinates of $W$ are $(-5\sqrt{3}, -5)$.

### Problems

**25.** Judging from the figure:

- The curve looks the same from $t = -24$ to $t = 0$ as from $t = 0$ to $t = 24$ and as from $t = 24$ to $t = 48$, so it repeats with a period of 24.
- The midline is the dashed horizontal line $y = -500$.
- The vertical distance from the first peak to the midline is 2000, so the amplitude is 2000.

**29.** $g(x) = \cos x$, $a = 90°$ and $b = 1$.

**33.** $f(x) = \sin(x + 90°)$
$g(x) = \sin(x - 90°)$

**37.** Since the wheel has diameter 4.5 meters, it has radius 2.25 meters. Half of the bucket dips below the water surface in its lowest position, so the center of the bucket is $2.25 + 0.25 = 2.5$ meters away from the center of the water wheel. The lowest height of the center of the bucket, and thus the lowest function value, will be 0 when $\theta = 270°$. The highest position of the center of the bucket will be 5 m above the water when $\theta = 90°$. Therefore the height of the center of the water bucket above the river is given by a sinusoidal function with midline 2.5 and amplitude 2.5: $h(\theta) = 2.5 + 2.5 \sin \theta$.

# Solutions for Section 7.4

## Exercises

**1.** $\sin 0° = 0$, $\cos 0° = 1$, $\tan 0° = \sin 0° / \cos 0° = 0/1 = 0$.

**5.** By the Pythagorean Theorem, we know that the third side must be $\sqrt{7^2 - 2^2} = \sqrt{45}$.

   **(a)** Since $\sin \theta$ is opposite side over hypotenuse, we have $\sin \theta = \sqrt{45}/7$.
   **(b)** Since $\cos \theta$ is adjacent side over hypotenuse, we have $\cos \theta = 2/7$.
   **(c)** Since $\tan \theta$ is opposite side over adjacent side, we have $\tan \theta = \sqrt{45}/2$.

**9.** By the Pythagorean Theorem, we know that the third side must be $\sqrt{11^2 - 2^2} = \sqrt{117}$.

   **(a)** Since $\sin \theta$ is opposite side over hypotenuse, we have $\sin \theta = \sqrt{117}/11$.
   **(b)** Since $\cos \theta$ is adjacent side over hypotenuse, we have $\cos \theta = 2/11$.
   **(c)** Since $\tan \theta$ is opposite side over adjacent side, we have $\tan \theta = \sqrt{117}/2$.

**13.** Since $\cos 37° = 6/r$, we have $r = 6/\cos 37°$. Similarly, since $\tan 37° = q/6$, we have $q = 6 \tan 37°$.

**17.** $\cos 90° = 0$

**21.** Since $225°$ is in the third quadrant,
$$\tan 225° = \tan 45° = 1.$$

## Problems

**25.** Figure 7.6 illustrates this situation.

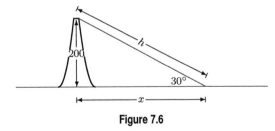

**Figure 7.6**

We have a right triangle with legs $x$ and 200 and hypotenuse $h$. Thus,
$$\sin 30° = \frac{200}{h}$$
$$h = \frac{200}{\sin 30°} = \frac{200}{0.5} = 400 \text{ feet.}$$

To find the distance $x$, we can relate the angle and its opposite and adjacent legs by writing

$$\tan 30° = \frac{200}{x}$$

$$x = \frac{200}{\tan 30°} \approx 346.410 \text{ feet.}$$

We could also write the equation $x^2 + 200^2 = h^2$ and substitute $h = 400$ ft to solve for $x$.

29. Since the distance from $P$ to $A$ is $\dfrac{50}{\tan 42°}$ and the distance from $P$ to $B$ is $\dfrac{50}{\tan 35°}$,

$$d = \frac{50}{\tan 35°} - \frac{50}{\tan 42°} \approx 15.877 \text{ feet.}$$

# Solutions for Section 7.5

## Exercises

1. Since we are looking for the angle $\theta$, we have $\sin \theta = 0.876$, so $\theta = \sin^{-1} 0.876 = 61.164°$.

5. Since the output of the cosine function is $0 < \cos \theta < 1$, there is no angle whose cosine is 2.614.

9. We know that $\sin 60° = \dfrac{\sqrt{3}}{2}$. Therefore, $\theta = 60°$.

13. We know that $\sin 45° = \dfrac{\sqrt{2}}{2}$. Therefore, $\theta = 45°$.

17. We have

$$B = 90° - A = 62°$$
$$a = c \cdot \sin A = 20 \sin 28° = 9.389$$
$$b = c \cdot \sin B = 20 \sin 62° = 17.659.$$

21. Since $(\tan c)^{-1} = \dfrac{1}{\tan c} = d$, we have $\tan c = \dfrac{1}{d}$. Therefore the angle is $c$ and the value is $\dfrac{1}{d}$.

## Problems

25. (a) We are looking for the value of the sine of $\left(\dfrac{1}{2}\right)°$. Using a calculator or computer, we have $\sin\left(\dfrac{1}{2}\right)° = 0.009$.

(b) We are looking for the angle in a right triangle whose sine is $\dfrac{1}{2}$. Therefore, we have $\sin^{-1}\left(\dfrac{1}{2}\right) = 30°$.

(c) We are looking for the reciprocal of the sine of $\left(\dfrac{1}{2}\right)°$. We have

$$(\sin x)^{-1} = \left(\sin \frac{1}{2}\right)^{-1}$$
$$= \frac{1}{\sin \frac{1}{2}}$$
$$= 114.593.$$

**29.** We have

$$6\cos\theta - 2 = 3$$
$$6\cos\theta = 5$$
$$\cos\theta = \left(\frac{5}{6}\right)$$
$$\theta = \cos^{-1}\left(\frac{5}{6}\right)$$
$$\theta = 33.557°.$$

**33.** We have

$$9\tan(5\theta) + 1 = 10$$
$$9\tan(5\theta) = 9$$
$$\tan(5\theta) = 1$$
$$5\theta = \tan^{-1}(1)$$
$$5\theta = 45°$$
$$\theta = 9°.$$

**37.** We have

$$5\tan(4\theta) + 4 = 2(\tan(4\theta) + 5)$$
$$5\tan(4\theta) + 4 = 2\tan(4\theta) + 10$$
$$3\tan(4\theta) = 6$$
$$\tan(4\theta) = 2$$
$$4\theta = \tan^{-1}(2)$$
$$4\theta = 63.435°$$
$$\theta = 15.859°.$$

**41.** See Figure 7.7. The angle $\theta$ is the sun's angle of elevation. Here, $\tan\theta = \dfrac{50}{60} = \dfrac{5}{6}$. So, $\theta = \tan^{-1}\left(\dfrac{5}{6}\right) \approx 39.806°$.

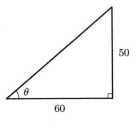

**Figure 7.7**

## Solutions for Section 7.6

### Exercises

**1.** By the Law of Sines, we have

$$\frac{x}{\sin 100^\circ} = \frac{6}{\sin 18^\circ}$$

$$x = 6 \left( \frac{\sin 100^\circ}{\sin 18^\circ} \right) \approx 19.121.$$

**5.** In Figure 7.8, we have

$\theta = 180^\circ - 90^\circ - 10^\circ$        $a = 12 \cos 10^\circ$        $b = 12 \sin 10^\circ$

$\theta = 80^\circ.$        $a \approx 12(0.985)$        $b \approx 12(0.174)$

    $a \approx 11.818.$        $b \approx 2.084.$

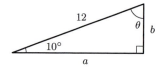

**Figure 7.8**

**9.** Using the Law of Cosines, we have

$$41^2 = 20^2 + 28^2 - 2 \cdot 20 \cdot 28 \cos C$$

$$497 = -1120 \cos C$$

$$C = \cos^{-1} \left( -\frac{497}{1120} \right)$$

$$C = 116.343^\circ.$$

Using the Law of Cosines again (though we could use the Law of Sines), we have

$$20^2 = 41^2 + 28^2 - 2 \cdot 41 \cdot 28 \cos A$$

$$-2065 = -2296 \cos A$$

$$A = \cos^{-1} \left( \frac{2065}{2296} \right)$$

$$A = 25.922^\circ.$$

Thus, $B = 180 - A - C = 37.735^\circ$.

**13.** We begin by using the law of cosines to find side $c$:

$$c^2 = 9^2 + 8^2 - 2 \cdot 9 \cdot 8 \cos 80^\circ$$

$$c^2 = 119.995$$

$$c = 10.954.$$

We can now use the law of sines to find the other two angles.

$$\frac{\sin B}{8} = \frac{\sin 80^\circ}{10.954}$$

$$\sin B = 8\frac{\sin 80°}{10.954}$$
$$B = \sin^{-1} 0.719$$
$$B = 45.990°.$$

Therefore, $A = 180° - 80° - 45.990 = 54.010°$.

**17.** We begin by finding the angle $C$, which is $180° - 105° - 9° = 66°$.
We can now use the law of sines to find the other two sides.

$$\frac{b}{\sin 9°} = \frac{15}{\sin 66°}$$
$$b = \sin 9° \cdot \frac{15}{\sin 66°}$$
$$b = 2.569.$$

Similarly,

$$\frac{a}{\sin 105°} = \frac{15}{\sin 66°}$$
$$a = \sin 105° \cdot \frac{15}{\sin 66°}$$
$$a = 15.860.$$

**21.** We begin by finding the angle $B$, which is $180° - 150° - 12° = 18°$.
We can now use the law of sines to find the other two sides.

$$\frac{a}{\sin 12°} = \frac{5}{\sin 150°}$$
$$a = \sin 12° \cdot \frac{5}{\sin 150°}$$
$$a = 2.079.$$

Similarly,

$$\frac{b}{\sin 18°} = \frac{5}{\sin 150°}$$
$$b = \sin 18° \cdot \frac{5}{\sin 150°}$$
$$b = 3.090.$$

**25.** First, we recognize that it is possible that there are two triangles, since we may have the ambiguous case. However, we know that angle $C$ must be less than $72°$, since the side across from it is shorter than 13. Thus, we begin by finding the angle $C$ using the law of sines:

$$\frac{\sin C}{4} = \frac{\sin 72°}{13}$$
$$\sin C = 4 \cdot \frac{\sin 72°}{13}$$
$$C = \sin^{-1} 0.293$$
$$C = 17.016°.$$

We can now solve for $A$, which is $180° - 72° - 17.016° = 90.984°$.

Using the law of sines, we can solve for side $a$:

$$\frac{a}{\sin 90.984} = \frac{13}{\sin 72°}$$

$$a = \sin 90.984 \cdot \frac{13}{\sin 72°}$$

$$a = 13.667.$$

29.

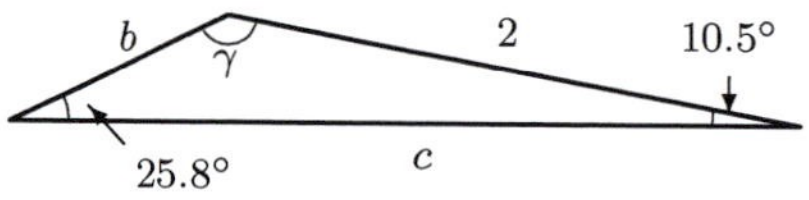

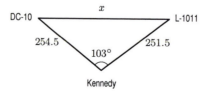

**Figure 7.9**

The Law of Sines tells us that $\frac{b}{\sin 10.5°} = \frac{2}{\sin 25.8°}$, so $b = 0.837$ m. We have $\gamma = 180° - 10.5° - 25.8° = 143.7°$. We use this value to find side length $c$. We have $c^2 = 2^2 + (.837)^2 - 2(2)(.837)\cos 143.7° \approx 7.401$, or $c = 2.720$ m.

## Problems

33. **(a)** By the Law of Sines, we have

$$\frac{\sin\theta}{3} = \frac{\sin 110°}{10}$$

$$\sin\theta = \left(\frac{3}{10}\right)\sin 110° \approx 0.282.$$

**(b)** If $\sin\theta = 0.282$, then $\theta \approx 16.374°$ (as found on a calculator) or $\theta \approx 180° - 16.374° \approx 163.626°$. Since the triangle already has a $110°$ angle, $\theta \approx 16.374°$. (The $163.626°$ angle would be too large.)

**(c)** The height of the triangle is $10\sin\theta = 10 \cdot 0.282 = 2.819$ cm. Since the sum of the angles of a triangle is $180°$, and we know two of the angles, $\theta = 16.374°$ and $110°$, so the third angle is $180° - 16.374° - 110° = 53.626°$. By the Law of of Sines, we have

$$\frac{\sin 110°}{10} = \frac{\sin 53.626°}{\text{Base}}$$

$$\text{Base} = \frac{10\sin 53.626°}{\sin 110°} \approx 8.568 \text{ cm}.$$

Thus, the triangle has

$$\text{Area} = \frac{1}{2}\text{Base} \cdot \text{Height} = \frac{1}{2} \cdot 8.568 \cdot 2.819 = 12.077 \text{ cm}^2.$$

37. After half an hour, the DC-10 has traveled $509 \cdot (1/2) = 254.5$ miles and the L-1011 has traveled $503 \cdot (1/2) = 251.5$ miles. See Figure 7.10. Using the Law of Cosines:

$$x^2 = 254.5^2 + 251.5^2 - 2(254.5)(251.5)\cos 103°$$

$$x = 396.004 \text{ miles.}$$

**Figure 7.10**

**41.** If we look at Figure 7.11, we see that there are two triangles: the original triangle with angles $A$, $B$, $C$ and the right triangle with hypotenuse $b$.

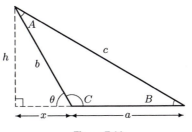

**Figure 7.11**

The Pythagorean theorem gives

$$x^2 + h^2 = b^2,$$

or

$$h^2 = b^2 - x^2.$$

If we apply the Pythagorean theorem to the right triangle with legs $h$ and $x + a$ we obtain

$$(x + a)^2 + h^2 = c^2.$$

Substituting $h^2 = b^2 - x^2$ into this equation gives

$$x^2 + 2ax + a^2 + \underbrace{b^2 - x^2}_{h^2} = c^2.$$

This, in turn, reduces to

$$a^2 + b^2 + 2ax = c^2.$$

We now determine $x$. Since $C$ is obtuse, $\cos C$ will be negative. We have

$$\cos C = -\cos\theta = -\frac{x}{b},$$

or

$$x = -b\cos C.$$

Substituting this expression for $x$ into our equation gives the Law of Cosines:

$$a^2 + b^2 - 2a\underbrace{b\cos C}_{x} = c^2.$$

**45.** The three stakes are at $A$, $B$, $C$ in Figure 7.12. Using the Law of Cosines, we have

$$x^2 = 82^2 + 97^2 - 2(82)(97)\cos 125°$$
$$x \approx 158.926.$$

The mound is approximately 158.926 feet wide.

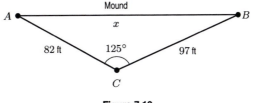

**Figure 7.12**

## Solutions for Chapter 7 Review

### Exercises

**1.** This function appears to be periodic because it repeats regularly.

**5.** This function does not appear to be periodic. Though it does rise and fall, it does not do so regularly.

**9.**

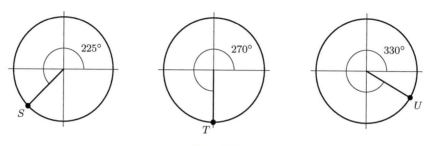

**Figure 7.13**

$$S = (-0.707, -0.707), T = (0, -1), U = (0.866, -0.5)$$

**13.**

$$x = r \cos \theta = 16 \cos(-72°) \approx 4.944$$

and

$$y = r \sin \theta = 16 \sin(-72°) \approx -15.217,$$

so the approximate coordinates of $Z$ are $(4.944, -15.217)$.

**17.** Since we are looking for the angle $\theta$, we have $\tan \theta = \frac{5}{3}$, so $\theta = \tan^{-1} \frac{5}{3} = 59.036°$.

**21.** We know that $\cos 45° = \frac{\sqrt{2}}{2}$. Therefore, $\theta = 45°$.

**25.** Since $\tan^{-1}\left(c^{-1}\right) = d$ means $\tan d = \frac{1}{c}$, the angle is $d$ and the value is $\frac{1}{c}$.

### Problems

**29.** In graph $A$, the average speed is relatively high with little variation, which corresponds to (iii). In graph $B$, the average speed is lower and there are significant speed-ups and slow-downs, which corresponds to (i). In graph $C$, the average speed is low and frequently drops to 0, which this corresponds to (ii).

**33.** By plotting the data in Figure 7.14, we can see that the midline is at $h = 2$ (approximately). Since the maximum value is 3 and the minimum value is 1, we have

$$\text{Amplitude} = 2 - 1 = 1.$$

Finally, we can see from the graph that one cycle has been completed from time $t = 0$ to time $t = 1$, so the period is 1 second.

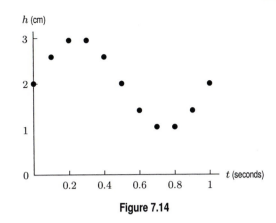

**Figure 7.14**

**37.** We know all three sides of this triangle, but only one of its angles. We find the value of $\sin \theta$ and $\sin \phi$ in this right triangle:

$$\sin \theta = \frac{\text{opposite}}{\text{hypotenuse}} = \frac{3}{5} = 0.6$$

and

$$\sin \phi = \frac{\text{opposite}}{\text{hypotenuse}} = \frac{4}{5} = 0.8.$$

Using inverse sines, we know that if $\sin \phi = 0.8$, then $\phi = \sin^{-1}(0.8) \approx 53.130°$. Similarly $\sin \theta = 0.6$ means $\theta = \sin^{-1}(0.6) \approx 36.870°$. Notice $\phi + \theta = 90°$, which has to be true in a right triangle.

**41.** First check to see if there is a right triangle. It is not because $25^2 + 52^2 \neq 63^2$. So we must use the Law of Cosines:

$$25^2 = 63^2 + 52^2 - 2(63)(52) \cos \theta$$

$$\frac{25^2 - 63^2 - 52^2}{-2(63)(52)} = \cos \theta$$

$$0.9231 \approx \cos \theta$$

$$\arccos(0.9231) \approx \theta$$

$$22.620° \approx \theta$$

**45.** The angle of observation is labeled $\theta$ in Figure 7.92. The distance 200 meters forms the adjacent side for this angle and the height of the balloon, $h$, is the opposite side. Thus,

$$\tan \theta = \frac{h}{200},$$

so

$$h = 200 \tan \theta.$$

## CHECK YOUR UNDERSTANDING

**1.** True. The point $(1, 0)$ on the unit circle is the starting point to measure angles so $\theta = 0°$.

**5.** True. Because $P$ and $Q$ have the same $x$-coordinates.

**9.** False. The angle $315°$ is in the fourth quadrant, so the cosine value is positive. The correct value is $\sqrt{2}/2$.

**13.** False. A unit circle must have a radius of 1.

**17.** False. Both acute angles are 45 degrees, and $\sin 45° = \sqrt{2}/2$.

**21.** True. By the Law of Cosines, we have $p^2 = n^2 + r^2 - 2nr \cos P$, so $\cos P = (n^2 + r^2 - p^2)/(2nr)$.

**25.** True. Identify the opposite angles as $B$ and $L$ and use the Law of Sines to obtain $\frac{LA}{\sin B} = \frac{BA}{\sin L}$. Thus $\frac{LA}{BA} = \frac{\sin B}{\sin L}$.

## Solutions to Skills for Chapter 7

### Exercises

**1.** We have $\sin 30° = 1/2$.

**5.** Since we know that the $x$-coordinate on the unit circle at $-60°$ is the same as the $x$-coordinate at $60°$, we know that $\cos(-60°) = \cos 60° = 1/2$.

**9.** Since $135°$ is in the second quadrant,

$$\sin 135° = \sin 45° = \frac{1}{\sqrt{2}}.$$

**13.** Since $405°$ is in the first quadrant,

$$\sin 405° = \sin 45° = \frac{1}{\sqrt{2}}.$$

**17.** The reference angle for $300°$ is $360° - 300° = 60°$, so $\sin 300° = -\sin 60° = -\sqrt{3}/2$.

### Problems

**21.** Since $\sin 30° = x/10$, we have $x = 10(1/2) = 5$.

**25.** In a $45°$-$45°$-$90°$ triangle, the two legs are equal and the hypotenuse is $\sqrt{2}$ times the length of a leg. So the sides are 5 and 5 and $5\sqrt{2}$.

**29.** A right triangle with two equal sides is a $45°$-$45°$-$90°$ triangle. In such triangles the length of the hypotenuse side is $\sqrt{2}$ times the length of a leg, so the third side is $4\sqrt{2}$.

# CHAPTER EIGHT

## Solutions for Section 8.1

### Exercises

**1.** To convert $60°$ to radians, multiply by $\pi/180°$:

$$60° \left( \frac{\pi}{180°} \right) = \left( \frac{60°}{180°} \right) \pi = \frac{\pi}{3}.$$

We say that the radian measure of a $60°$ angle is $\pi/3$.

**5.** In order to change from degrees to radians, we multiply the number of degrees by $\pi/180$, so we have $150 \cdot \pi/180$, giving $\frac{5}{6}\pi$ radians.

**9.** In order to change from radians to degrees, we multiply the number of radians by $180/\pi$, so we have $\frac{7}{2}\pi \cdot 180/\pi$, giving 630 degrees.

**13.** In order to change from radians to degrees, we multiply the number of radians by $180/\pi$, so we have $45 \cdot 180/\pi$, giving $8100/\pi \approx 2578.310$ degrees.

**17.** If we go around twice, we make two full circles, which is $2\pi \cdot 2 = 4\pi$ radians. Since we're going around in the negative direction, we have $-4\pi$ radians.

**21.** The arc length, $s$, corresponding to an angle of $\theta$ radians in a circle of radius $r$ is $s = r\theta$. In order to change from degrees to radians, we multiply the number of degrees by $\pi/180$, so we have $45 \cdot \pi/180$, giving $\frac{\pi}{4}$ radians. Thus, our arc length is $6.2\pi/4 \approx 4.869$.

**25.** (a) The reference angle for $2\pi/3$ is $\pi - 2\pi/3 = \pi/3$, so $\sin(2\pi/3) = \sin(\pi/3) = \sqrt{3}/2$.
 (b) The reference angle for $3\pi/4$ is $\pi - 3\pi/4 = \pi/4$, so $\cos(3\pi/4) = -\cos(\pi/4) = -1/\sqrt{2}$.
 (c) The reference angle for $-3\pi/4$ is $\pi - 3\pi/4 = \pi/4$, so $\tan(-3\pi/4) = \tan(\pi/4) = 1$.
 (d) The reference angle for $11\pi/6$ is $2\pi - 11\pi/6 = \pi/6$, so $\cos(11\pi/6) = \cos(\pi/6) = \sqrt{3}/2$.

### Problems

**29.**

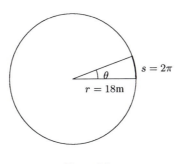

**Figure 8.1**

In Figure 8.1, we have $s = 2\pi$ and $r = 18$. Therefore,

$$\theta = \frac{s}{r} = \frac{2\pi}{18} = \frac{\pi}{9}.$$

Now,

$$\frac{\pi}{9} \text{ radians} = \frac{\pi}{9}\left(\frac{180°}{\pi}\right) = 20°.$$

Therefore, an arc of length $2\pi$ m on a circle of radius 18 m determines an angle of $\pi/9$ radians or $20°$.

**33.** We have $\theta = 1.3$ rad or, in degrees,

$$1.3\left(\frac{180°}{\pi}\right) = 74.4845°.$$

We also have $r = 12$, so

$$s = 12(1.3) = 15.6,$$

and $P = (r\cos\theta, r\sin\theta) = (3.2100, 11.5627)$.

**37.** (a) Negative
   (b) Negative
   (c) Positive
   (d) Positive

**41.** Since the ant traveled three units on the unit circle, the traversed arc must be spanned by an angle of three radians. Thus the ant's coordinates must be

$$(\cos 3, \sin 3) \approx (-0.99, 0.14).$$

**45.** (a) 1 radian is $180/\pi$ degrees so 30 radians is

$$30 \cdot \frac{180°}{\pi} \approx 1718.873°.$$

To check this answer, divide $1718.873°$ by $360°$ to find this is roughly 5 revolutions. A revolution in radians has a measure of $2\pi \approx 6$, so $5 \cdot 6 = 30$ radians makes sense.
   (b) 1 degree is $\pi/180$ radians, so $\pi/6$ degrees is

$$\frac{\pi}{6} \cdot \frac{\pi}{180} = \frac{\pi^2}{6 \cdot 180} \approx 0.00914 \text{ radians}.$$

This makes sense because $\pi/6$ is about $1/2$, and $1/2$ a degree is very small. One radian is about $60°$ so $\frac{1}{2}°$ is a very small part of a radian.

**49.** The value of $t$ is bigger than the value of $\sin t$ on $0 < t < \pi/2$. On a unit circle, the vertical segment, $\sin t$, is shorter than the arc, $t$. See Figure 8.2.

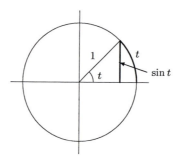

**Figure 8.2**

## Solutions for Section 8.2

### Exercises

**1.** The midline is $y = 0$. The amplitude is 6. The period is $2\pi$.

**5.** We see that the phase shift is $-4$, since the function is in a form that shows it. To find the horizontal shift, we factor out a 3 within the cosine function, giving us

$$y = 2\cos\left(3\left(t + \frac{4}{3}\right)\right) - 5.$$

Thus, the horizontal shift is $-4/3$.

**9.** This function resembles a sine curve in that it passes through the origin and then proceeds to grow from there. We know that the smallest value it attains is $-4$ and the largest it attains is 4, thus its amplitude is 4. It has a period of 1. Thus in the equation

$$g(t) = A\sin(Bt)$$

we know that $A =$ and

$$1 = \text{period} = \frac{2\pi}{B}.$$

So $B = 2\pi$, and then

$$h(t) = 4\sin(2\pi t).$$

**13.** The midline is $y = 4000$. The amplitude is $8000 - 4000 = 4000$. The period is 60, so the angular frequency is $2\pi/60$. The graph at $x = 0$ rises from its midline, so we use the sine. Thus,

$$y = 4000 + 4000\sin\left(\frac{2\pi}{60}x\right).$$

### Problems

**17.** See Figure 8.3.

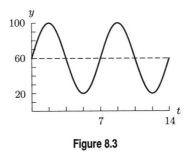

**Figure 8.3**

**21.** Because the period of $\sin x$ is $2\pi$, and the period of $\sin 2x$ is $\pi$, so from the figure in the problem we see that

$$f(x) = \sin x.$$

The points on the graph are $a = \pi/2$, $b = \pi$, $c = 3\pi/2$, $d = 2\pi$, and $e = 1$.

**25.** First, we note that the graph indicates that $f$ completes one cycle in 10 units. Therefore, the period of $f$ equals 10, and we can also see that by shifting the graph of $f$ 3 units to the right, we obtain the graph of $g$. Thus, $f$ must be shifted by $3/10$ of a period to the right to obtain $g$, meaning that the phase shift of $g$ is $2\pi \cdot (3/10) = 3\pi/5$. Therefore, we have $g(x) = 10\sin((\pi/5)x - 3\pi/5)$.

**29.** $f(t) = 20 + 15\sin\left(\dfrac{\pi}{2}t + \dfrac{\pi}{2}\right)$

**33. (a)** The midline is at $P = (2200 + 1300)/2 = 1750$. The amplitude is $|A| = 2200 - 1750 = 450$. The population starts at its minimum so it is modeled by vertically reflected cosine curve. This means $A = -450$, and that there is no phase shift. Since the period is 12, we have $B = 2\pi/12 = \pi/6$. This means the formula is

$$P = f(t) = -450\cos\left(\frac{\pi}{6}t\right) + 1750.$$

**(b)** The midline, $P = 1750$, is the average population value over one year. The period is 12 months (or 1 year), which means the cycle repeats annually. The amplitude is the amount that the population varies above and below the average annual population.

**(c)** Figure 8.4 is a graph of $f(t) = -450\cos(\frac{\pi}{6}t) + 1750$ and $P = 1500$.

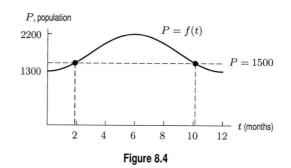

**Figure 8.4**

From a graph we get approximations of $t_1 \approx 1.9$ and $t_2 \approx 10.1$. This means that the population is 1500 sometime in late February and again sometime in early November.

**37.** This function has an amplitude of 1 and a period of 0.5, and resembles an inverted sine graph. Thus $y = -f(2x)$.

**41.** Figure 8.5 highlights the two parts of the graph. In the first hour, the plane is approaching Boston. In the second hour, the plane is circling Boston.

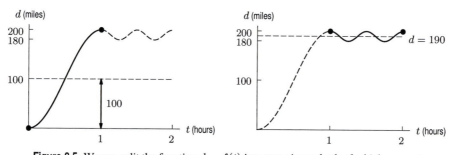

**Figure 8.5:** We can split the function $d = f(t)$ into two pieces, both of which are cosine curves

From Figure 8.5, we see that both parts of $f(t)$ look like cosine curves. The first part has the equation

$$f(t) = -100\cos(\pi t) + 100, \quad \text{for } 0 \le t \le 1.$$

In the second part, the period is 1/2, the midline is 190, and the amplitude is $200 - 190 = 10$, so

$$f(t) = 10\cos(4\pi t) + 190, \quad \text{for } 1 \le t \le 2.$$

Thus, a piecewise formula for $f(t)$ could be

$$f(t) = \begin{cases} -100\cos(\pi t) + 100 & \text{for } 0 \le t \le 1 \\ 10\cos(4\pi t) + 190 & \text{for } 1 < t \le 2. \end{cases}$$

**45. (a)** Although the graph has a rough wavelike pattern, the wave is not perfectly regular in each 7-day interval. A true periodic function has a graph which is absolutely regular, with values that repeat exactly every period.

**(b)** Usage spikes every 7 days or so, usually about midweek (8/7, 8/14, 8/21, etc.). It drops to a low point every 7 days or so, usually on Saturday or Sunday (8/10, 8/17, 8/25, etc.). This indicates that scientists use the site less frequently on weekends and more frequently during the week.

**(c)** See Figure 8.6 for one possible approximation. The function shown here is given by $n = a\cos(B(t-h)) + k$ where $t$ is the number of days from Monday, August 5, and $a = 45,000$, $B = 2\pi/7$, $h = 2$, and $k = 100,000$. The midline $k = 100,000$ tells us that usage rises and falls around an approximate average of 100,000 connections per day. The amplitude $a = 45,000$ tells us that usage tends to rise or fall by about 45,000 from the average over the course of the week. The period is 7 days, or one week, giving $B = 2\pi/7$, and the curve resembles a cosine function shifted to the right by about $h = 2$ days. Thus,

$$n = 45,000\cos\left(\frac{2\pi}{7}(t-2)\right) + 100,000.$$

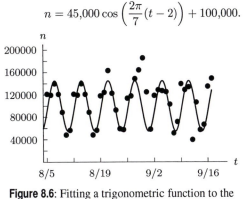

**Figure 8.6**: Fitting a trigonometric function to the arXiv.org usage data

# Solutions for Section 8.3

## Exercises

**1.** 1

**5.** $\dfrac{-1}{\sqrt{3}}$

**9.** Since $\sec(-\pi/6) = 1/\cos(-\pi/6)$, we know that $\sec(-\pi/6) = 1/(\sqrt{3}/2) = 2/\sqrt{3}$.

**13.** Writing $\sec t = 1/\cos t$, we have
$$\sec t \cos t = 1.$$

**17.** Expanding the square and combining terms gives
$$(\sin x - \cos x)^2 + 2\sin x \cos x = \sin^2 x - 2\sin x \cos x + \cos^2 + 2\sin x \cos x = \sin^2 x + \cos^2 x = 1.$$

## Problems

**21.** Since $\sec\theta = 1/\cos\theta$, we have $\sec\theta = 1/(1/2) = 2$. Since $1 + \tan^2\theta = \sec^2\theta$,
$$1 + \tan^2\theta = 2^2$$
$$\tan^2\theta = 4 - 1$$
$$\tan\theta = \pm\sqrt{3}.$$
Since $0 \le \theta \le \pi/2$, we know that $\tan\theta \ge 0$, so $\tan\theta = \sqrt{3}$.

**25.** This looks like a tangent graph. At $\pi/4$, $\tan\theta = 1$. Since on this graph, $f(\pi/4) = 1/2$, and since it appears to have the same period as $\tan\theta$ without a horizontal or vertical shift, a possible formula is $f(\theta) = \frac{1}{2}\tan\theta$.

**29. (a)** $\sin^2\phi = 1 - \cos^2\phi = 1 - (0.4626)^2$ and $\sin\phi$ is negative, so $\sin\phi = -\sqrt{1 - (0.4626)^2} = -0.8866$. Thus $\tan\phi = (\sin\phi)/(\cos\phi) = (-0.8866)/(0.4626) = -1.9166$.

 **(b)** $\cos^2\theta = 1 - \sin^2\theta = 1 - (-0.5917)^2$ and $\cos\theta$ is negative, so $\cos\theta = -\sqrt{1 - (-0.5917)^2} = -0.8062$. Thus $\tan\theta = (\sin\theta)/(\cos\theta) = (-0.5917)/(-0.8062) = 0.7339$.

**33.** First notice that $\cos\theta = \frac{x}{2}$, then $\sin^2\theta = 1 - \cos^2\theta = 1 - (x/2)^2 = 1 - x^2/4 = (4 - x^2)/4$, so $\sin\theta = \sqrt{(4 - x^2)/4} = \sqrt{4 - x^2}/2$. Thus $\tan\theta = \sin\theta/\cos\theta = \sqrt{4 - x^2}/2 \cdot 2/x = \sqrt{4 - x^2}/x$.

**37. (a)** The graph of $y = \tan t$ has vertical asymptotes at odd multiples of $\pi/2$, that is, at $\pi/2$, $3\pi/2$, $5\pi/2$, etc., and their negatives. The graph of $y = \cos t$ has $t$-intercepts at the same values.

 **(b)** The graph of $y = \tan t$ has $t$-intercepts at multiples of $\pi$, that is, at $0$, $\pm\pi$, $\pm 2\pi$, $\pm 3\pi$, etc. The graph of $y = \sin t$ has $t$-intercepts at the same values.

# Solutions for Section 8.4

## Exercises

**1.** We use the inverse tangent function on a calculator to get $\theta = 1.570$.

**5.** We use the inverse tangent function on a calculator to get $5\theta + 7 = -0.236$. Solving for $\theta$, we get $\theta = -1.447$.

**9.** Graph $y = \cos t$ on $0 \le t \le 2\pi$ and locate the two points with $y$-coordinate $-0.24$. The $t$-coordinates of these points are approximately $t = 1.813$ and $t = 4.473$. See Figure 8.7.

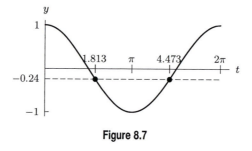

**Figure 8.7**

## Problems

**13.** We know that one solution to the equation $\cos t = 0.4$ is $t = \cos^{-1}(0.4) = 1.159$. From Figure 8.8, we see by the symmetry of the unit circle that another solution is $t = -1.159$. We know that additional solutions are given each time the angle $t$ wraps around the circle in either direction. This means that

$$1.159 + 1 \cdot 2\pi = 7.442 \qquad \text{wrap once around circle}$$
$$1.159 + 2 \cdot 2\pi = 13.725 \qquad \text{wrap twice around circle}$$
$$1.159 + (-1) \cdot 2\pi = -5.124 \qquad \text{wrap once around the other way}$$

and

$$-1.159 + 1 \cdot 2\pi = 5.124 \qquad \text{wrap once around circle}$$
$$-1.159 + 2 \cdot 2\pi = 11.407 \qquad \text{wrap once around circle}$$

$-1.159 + (-1) \cdot 2\pi = -7.442.$    wrap once around circle the other way

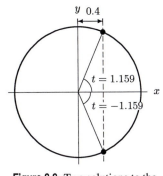

**Figure 8.8**: Two solutions to the
equation $\cos t = 0.4$

17. Since $\cos t = 1/2$, we have $t = \pi/3$ and $t = 5\pi/3$.

21. Since $\tan t = 0$, we have $t = 0$, $t = \pi$, and $t = 2\pi$.

25. Looking at the graph of $y = \sin(2x)$ in Figure 8.9, we see it crosses the line $y = 0.3$ four times between 0 and $2\pi$, so there will be four solutions. Since $2x = \sin^{-1}(0.3) = 0.30469$, one solution is

$$x = \frac{0.30469}{2} = 0.1523.$$

The period of $y = \sin(2x)$ is $\pi$, so the other solutions in $0 \le x \le 2\pi$ are

$$x = 0.1523 + \pi = 3.294$$
$$x = \pi/2 - 0.1523 = 1.418$$
$$x = 3\pi/2 - 0.1523 = 4.560.$$

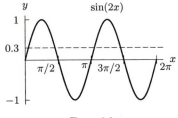

**Figure 8.9**

29. **(a)** Since $\cos(65°) = 0.4226$, a calculator set in degrees gives $\cos^{-1}(0.4226) = 65°$. We see in Figure 8.10 that all the angles with a cosine of 0.4226 correspond either to the point $P$ or to the point $Q$. We want solutions between $0°$ and $360°$, so $Q$ is represented by $360° - 65° = 295°$. Thus, the solutions are

$$\theta = 65°  \quad \text{and} \quad  \theta = 295°.$$

**(b)** A calculator gives $\tan^{-1}(2.145) = 65°$. Since $\tan\theta$ is positive in the first and third quadrants, the angles with a tangent of 2.145 correspond either to the point $P$ or the point $R$ in Figure 8.11. Since we are interested in solutions between $0°$ and $720°$, the solutions are

$$\theta = 65°, \quad 245°, \quad 65° + 360°, \quad 245° + 360°.$$

That is

$$\theta = 65°, \quad 245°, \quad 425°, \quad 605°.$$

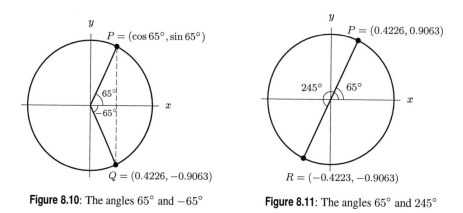

Figure 8.10: The angles $65°$ and $-65°$     Figure 8.11: The angles $65°$ and $245°$

**33.** By sketching a graph, we see that there are four solutions (see Figure 8.12). To find the four solutions, we begin by calculating $\sin^{-1}(-0.8) = -0.927$. Therefore, the length labeled "$b$" in Figure 8.12 is given by $b = -\sin^{-1}(-0.8) = 0.927$. Now, using the symmetry of the graph of the sine function, we can see that the four solutions are given by $x = \pi + b = 4.069$, $x = 2\pi - b = 5.356$, $x = 3\pi + b = 10.352$, and $x = 4\pi - b = 11.639$.

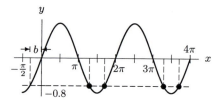

Figure 8.12

**37.** The angle between $-\pi/2$ and $0$ whose tangent is $-3$ is $\tan^{-1}(-3)$. Thus, the angle in the second quadrant is

$$\theta = \pi + \tan^{-1}(-3) \approx 1.893$$

**41.** From Figure 8.13, we see that $2\sin t \cos t - \cos t = 0$ has four roots between $0$ and $2\pi$. They are approximately $t = 0.52$, $t = 1.57$, $t = 2.62$, and $t = 4.71$.

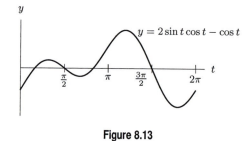

Figure 8.13

To solve the problem symbolically, we factor out $\cos t$:

$$2\sin t \cos t - \cos t = \cos t(2\sin t - 1) = 0.$$

So solutions occur either when $\cos t = 0$ or when $2 \sin t - 1 = 0$. The equation $\cos t = 0$ has solutions $\pi/2$ and $3\pi/2$. The equation $2 \sin t - 1 = 0$ has solution $t = \arcsin(1/2) = \pi/6$, and also $t = \pi - \pi/6 = 5\pi/6$. Thus the solutions to the original problem are

$$t = \frac{\pi}{2}, \frac{3\pi}{2}, \frac{\pi}{6} \text{ and } \frac{5\pi}{6}.$$

**45.** The curve is a sine curve with an amplitude of 5, a period of 8 and a vertical shift of $-3$. Thus the equation for the curve is $y = 5 \sin\left(\frac{\pi}{4}x\right) - 3$. Solving for $y = 0$, we have

$$5 \sin\left(\frac{\pi}{4}x\right) = 3$$
$$\sin\left(\frac{\pi}{4}x\right) = \frac{3}{5}$$
$$\frac{\pi}{4}x = \sin^{-1}\left(\frac{3}{5}\right)$$
$$x = \frac{4}{\pi} \sin^{-1}\left(\frac{3}{5}\right) \approx 0.819.$$

This is the $x$-coordinate of $P$. The $x$-coordinate of $Q$ is to the left of 4 by the same distance $P$ is to the right of $O$, by the symmetry of the sine curve. Therefore,

$$x \approx 4 - 0.819 = 3.181$$

is the $x$-coordinate of $Q$.

**49. (a)** Graph $y = 3 - 5 \sin 4t$ on the interval $0 \le t \le \pi/2$, and locate values where the function crosses the $t$-axis. Alternatively, we can find the points where the graph $5 \sin 4t$ and the line $y = 3$ intersect. By looking at the graphs of these two functions on the interval $0 \le t \le \pi/2$, we find that they intersect twice. By zooming in we can identify these points of intersection as roughly $t_1 \approx 0.16$ and $t_2 \approx 0.625$. See Figure 8.14.

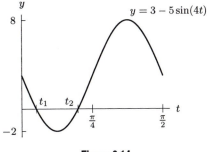

**Figure 8.14**

**(b)** Solve for $\sin(4t)$ and then use arcsine:

$$5 \sin(4t) = 3$$
$$\sin(4t) = \frac{3}{5}$$
$$4t = \arcsin\left(\frac{3}{5}\right).$$

So $t_1 = \dfrac{\arcsin(3/5)}{4} \approx 0.161$ is a solution. But the angle $\pi - \arcsin(3/5)$ has the same sine as $\arcsin(3/5)$. Solving $4t = \pi - \arcsin(3/5)$ gives $t_2 = \dfrac{\pi}{4} - \dfrac{\arcsin(3/5)}{4} \approx 0.625$ as a second solution.

**53. (a)** The value of $\sin t$ will be between $-1$ and $1$. This means that $k \sin t$ will be between $-k$ and $k$. Thus, $t^2 = k \sin t$ will be between $0$ and $k$. So

$$-\sqrt{k} \leq t \leq \sqrt{k}.$$

**(b)** Plotting $2 \sin t$ and $t^2$ on a calculator, we see that $t^2 = 2 \sin t$ for $t = 0$ and $t \approx 1.40$.

**(c)** Compare the graphs of $k \sin t$, a sine wave, and $t^2$, a parabola. As $k$ increases, the amplitude of the sine wave increases, and so the sine wave intersects the parabola in more points.

**(d)** Plotting $k \sin t$ and $t^2$ on a calculator for different values of $k$, we see that if $k \approx 20$, this equation will have a negative solution at $t \approx -4.3$, but that if $k$ is any smaller, there will be no negative solution.

## Solutions for Section 8.5

### Exercises

**1.** Quadrant IV.

**5.** Since $3.2\pi = 2\pi + 1.2\pi$, such a point is in Quadrant III.

**9.** Since $-7$ is an angle of $-7$ radians, corresponding to a rotation of just over $2\pi$, or one full revolution, in the clockwise direction, such a point is in Quadrant IV.

**13.** We have $180° < \theta < 270°$. See Figure 8.15.

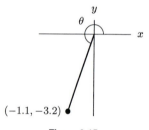

**Figure 8.15**

**17.** With $x = -\sqrt{3}$ and $y = 1$, find $r = \sqrt{(-\sqrt{3})^2 + 1^2} = \sqrt{4} = 2$. Find $\theta$ from $\tan \theta = y/x = 1/(-\sqrt{3})$. Thus, $\theta = \tan^{-1}(-1/\sqrt{3}) = -\pi/6$. Since $(-\sqrt{3}, 1)$ is in the second quadrant, $\theta = -\pi/6 + \pi = 5\pi/6$. The polar coordinates are $(2, 5\pi/6)$.

**21.** With $r = 2$ and $\theta = 5\pi/6$, we find $x = r \cos \theta = 2 \cos(5\pi/6) = 2(-\sqrt{3}/2) = -\sqrt{3}$ and $y = r \sin \theta = 2 \sin(5\pi/6) = 2(1/2) = 1$.
The rectangular coordinates are $(-\sqrt{3}, 1)$.

### Problems

**25.** Rewrite the left side using $\tan \theta = \dfrac{\sin \theta}{\cos \theta}$:

$$\frac{\sin \theta}{\cos \theta} = r \cos \theta - 2$$

In order to substitute $x = r \cos \theta$ and $y = r \sin \theta$, multiply by the $r$ in the numerator and denominator on the left

$$\frac{r \sin \theta}{r \cos \theta} = r \cos \theta - 2$$

and substitute:

$$\frac{y}{x} = x - 2.$$

So

$$y = x(x - 2)$$

or

$$y = x^2 - 2x.$$

**29.** By substituting $x = r\cos\theta$ and $y = r\sin\theta$, the equation becomes $2r\cos\theta \cdot r\sin\theta = 1$. This can be written as

$$r = \frac{1}{\sqrt{2\cos\theta\sin\theta}}.$$

**33.** Figure 8.16 shows that at 1 pm, we have:
In Cartesian coordinates, $M = (0, 4)$. In polar coordinates, $M = (4, \pi/2)$; that is $r = 4, \theta = \pi/2$.
At 1 pm, the hour hand points toward 1, so for $H$, we have $r = 3$ and $\theta = \pi/3$. Thus, the Cartesian coordinates of $H$ are

$$x = 3\cos\left(\frac{\pi}{3}\right) = 1.5, \qquad y = 3\sin\left(\frac{\pi}{3}\right) = \frac{3\sqrt{3}}{2} = 2.598.$$

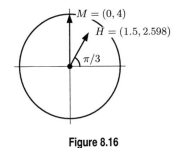

**Figure 8.16**

**37.** Figure 8.17 shows that at 9:15 am, the polar coordinates of the point $H$ (half-way between 9 and 9:30 on the clock face) are $r = 3$ and $\theta = 82.5° + 90° = 172.5\pi/180$. Thus, the Cartesian coordinates of $H$ are given by

$$x = 3\cos\left(\frac{172.5\pi}{180}\right) \approx -2.974, \quad y = 3\sin\left(\frac{172.5\pi}{180}\right) \approx 0.392.$$

In Cartesian coordinates, $H \approx (-2.974, 0.392)$. In polar coordinates, $H = (3, 172.5\pi/180)$. In Cartesian coordinates, $M = (4, 0)$. In polar coordinates, $M = (4, 0)$.

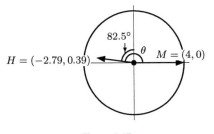

**Figure 8.17**

**41.** **(a)** Table 8.1 contains values of $r = 1 - \sin\theta$, both exact and rounded to one decimal.

**Table 8.1**

| $\theta$ | 0 | $\pi/3$ | $\pi/2$ | $2\pi/3$ | $\pi$ | $4\pi/3$ | $3\pi/2$ | $5\pi/3$ | $2\pi$ | $7\pi/3$ | $5\pi/2$ | $8\pi/3$ |
|---|---|---|---|---|---|---|---|---|---|---|---|---|
| $r$ | 1 | $1 - \sqrt{3}/2$ | 0 | $1 - \sqrt{3}/2$ | 1 | $1 + \sqrt{3}/2$ | 2 | $1 + \sqrt{3}/2$ | 1 | $1 - \sqrt{3}/2$ | 0 | $1 - \sqrt{3}/2$ |
| $r$ | 1 | 0.134 | 0 | 0.134 | 1 | 1.866 | 2 | 1.866 | 1 | 0.134 | 0 | 0.134 |

**(b)** See Figure 8.18.

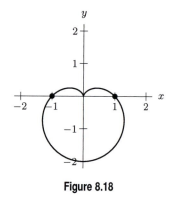

**Figure 8.18**

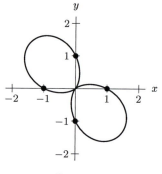

**Figure 8.19**

**(c)** The circle has equation $r = 1/2$. The cardioid is $r = 1 - \sin\theta$. Solving these two simultaneously gives

$$1/2 = 1 - \sin\theta,$$

or

$$\sin\theta = 1/2.$$

Thus, $\theta = \pi/6$ or $5\pi/6$. This gives the points $(x, y) = ((1/2)\cos\pi/6, (1/2)\sin\pi/6) = (\sqrt{3}/4, 1/4)$ and $(x, y) = ((1/2)\cos 5\pi/6, (1/2)\sin 5\pi/6) = (-\sqrt{3}/4, 1/4)$ as the location of intersection.

**(d)** The curve $r = 1 - \sin 2\theta$, pictured in Figure 8.19, has two regions instead of the one region that $r = 1 - \sin\theta$ has. This is because $1 - \sin 2\theta$ will be 0 twice for every $2\pi$ cycle in $\theta$, as opposed to once for every $2\pi$ cycle in $\theta$ for $1 - \sin\theta$.

**45.** See Figures 8.20 and 8.21. The first curve will be similar to the second curve, except the cardioid (heart) will be rotated clockwise by $90°$ ($\pi/2$ radians). This makes sense because of the identity $\sin\theta = \cos(\theta - \pi/2)$.

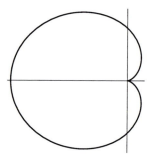

**Figure 8.20:** $r = 1 - \cos\theta$

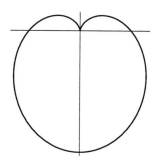

**Figure 8.21:** $r = 1 - \sin\theta$

# Solutions for Section 8.6

## Exercises

**1.** $5e^{i\pi}$

**5.** We have $(-3)^2 + (-4)^2 = 25$, and $\arctan(4/3) \approx 4.069$. So the number is $5e^{i4.069}$

**9.** $-3 - 4i$

**13.** We have $\sqrt{e^{i\pi/3}} = e^{(i\pi/3)/2} = e^{i\pi/6}$, thus $\cos\frac{\pi}{6} + i\sin\frac{\pi}{6} = \frac{\sqrt{3}}{2} + \frac{i}{2}$.

## Problems

**17.** One value of $\sqrt[3]{i}$ is $\sqrt[3]{e^{i\frac{\pi}{2}}} = (e^{i\frac{\pi}{2}})^{\frac{1}{3}} = e^{i\frac{\pi}{6}} = \cos\frac{\pi}{6} + i\sin\frac{\pi}{6} = \frac{\sqrt{3}}{2} + \frac{i}{2}$

**21.** One value of $(\sqrt{3} + i)^{1/2}$ is
$(2e^{i\frac{\pi}{6}})^{1/2} = \sqrt{2}e^{i\frac{\pi}{12}} = \sqrt{2}\cos\frac{\pi}{12} + i\sqrt{2}\sin\frac{\pi}{12} \approx 1.366 + 0.366i$

**25.** Substituting $A_2 = i - A_1$ into the second equation gives

$$iA_1 - (i - A_1) = 3,$$

so

$$iA_1 + A_1 = 3 + i$$
$$A_1 = \frac{3+i}{1+i} = \frac{3+i}{1+i} \cdot \frac{1-i}{1-i} = \frac{3 - 3i + i - i^2}{2}$$
$$= 2 - i$$

Therefore $A_2 = i - (2 - i) = -2 + 2i$.

**29.** Using Euler's formula, we have:

$$e^{i(2\theta)} = \cos 2\theta + i\sin 2\theta$$

On the other hand,

$$e^{i(2\theta)} = (e^{i\theta})^2 = (\cos\theta + i\sin\theta)^2 = (\cos^2\theta - \sin^2\theta) + i(2\cos\theta\sin\theta)$$

Equating real parts, we find

$$\cos 2\theta = \cos^2\theta - \sin^2\theta.$$

**33.** One polar form for $z = -8$ is $z = 8e^{i\pi}$ with $(r, \theta) = (8, \pi)$. Two more sets of polar coordinates for $z$ are $(8, 3\pi)$, and $(8, 5\pi)$. Three cube roots of $z$ are given by

$$\left(8e^{\pi i}\right)^{1/3} = 8^{1/3}e^{1/3 \cdot \pi i} = 2e^{\pi i/3} = 2\cos(\pi/3) + i2\sin(\pi/3)$$
$$= 1 + 1.732i.$$
$$\left(8e^{3\pi i}\right)^{1/3} = 8^{1/3}e^{1/3 \cdot 3\pi i} = 2e^{\pi i} = 2\cos\pi + i2\sin\pi$$
$$= -2.$$
$$\left(8e^{5\pi i}\right)^{1/3} = 8^{1/3}e^{1/3 \cdot 5\pi i} = 2e^{5\pi i/3} = 2\cos(5\pi/3) + i2\sin(5\pi/3)$$
$$= 1 - 1.732i.$$

**37.** By de Moivre's formula we have

$$(\cos 2\pi/3 + i\sin 2\pi/3)^3 = \cos(3 \cdot 2\pi/3) + i\sin(3 \cdot 2\pi/3) = 1 + i0 = 1.$$

## Solutions for Chapter 8 Review

### Exercises

**1.** In order to change from degrees to radians, we multiply the number of degrees by $\pi/180$, so we have $330 \cdot \pi/180$, giving $\frac{11}{6}\pi$ radians.

**5.** In order to change from radians to degrees, we multiply the number of radians by $180/\pi$, so we have $\frac{3}{2}\pi \cdot 180/\pi$, giving 270 degrees.

**9.** If we go around six times, we make six full circles, which is $2\pi \cdot 6 = 12\pi$ radians. Since we're going in the negative direction, we have $-12\pi$ radians.

**13.** Writing $\tan 2A = \sin 2A/\cos 2A$, we have

$$\frac{\sin 2A}{\tan 2A} + 2\cos 2A = \frac{\sin 2A}{\sin 2A/\cos 2A} + 2\cos 2A = 3\cos 2A.$$

**17.** **(a)** We are looking for the graph of a function with amplitude one but a period of $\pi$; only $C(t)$ qualifies.
   **(b)** We are looking for the graph of a function with amplitude one and period $2\pi$ but which is shifted up by two units; only $D(t)$ qualifies.
   **(c)** We are looking for the graph of a function with amplitude 2 and period $2\pi$; only $A(t)$ qualifies.
   **(d)** Only $B(t)$ is left and we are looking for the graph of a function with amplitude one and period $2\pi$ but which has been shifted to the left by two units. This checks with $B(t)$.

**21.** Since $-299°$ is almost $300°$, clockwise, such a point is in Quadrant I.

**25.** With $r = \pi/2$ and $\theta = 0$, we find $x = r\cos\theta = (\pi/2)\cos 0 = 1.571$ and $y = r\sin\theta = (\pi/2)\sin 0 = 0$.
   The rectangular coordinates are $(1.571, 0)$.

### Problems

**29.** The amplitude is 4, the period is $\frac{2\pi}{1} = 2\pi$, the phase shift and horizontal shift are 0. See Figure 8.22.

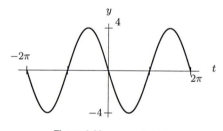

**Figure 8.22:** $y = -4\sin t$

**33.** The amplitude is 30, the midline is $y = 60$, and the period is 4.

**37.** We would use $y = \sin x$ for $f(x)$ and $k(x)$, as both these graphs cross the midline at $x = 0$. We would use $y = \cos x$ for $g(x)$ and $h(x)$, as both these graphs are as far as possible from the midline at $x = 0$.

**41.** See Figure 8.23.

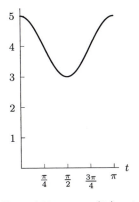

**Figure 8.23:** $y = \cos(2t) + 4$

**45.** **(a)** Since $B$ is the point $(1, 0)$, the circle has radius 1. Thus, $\sin\theta = \text{OE}$.

**(b)** From the definition of $\cos\theta$, we have $\cos\theta = \text{OA}$.

**(c)** In $\triangle\text{ODB}$, we have $OB = 1$. Since $\tan\theta = \text{Opp}/\text{Adj} = \text{DB}/1$, we have $\tan\theta = \text{DB}$.

**(d)** Let $P$ be the point of intersection of $\overline{FC}$ and $\overline{OD}$. Use the fact that $\triangle\text{OFP}$ and $\triangle\text{OEP}$ are similar, and form a proportion of the hypotenuse to the one unit side of the larger triangle.

$$\frac{\text{OF}}{1} = \frac{1}{\text{OE}} \qquad \text{or} \qquad \text{OF} = \frac{1}{\sin\theta}$$

**(e)** Use the fact that $\triangle\text{OCP}$ and $\triangle\text{OAP}$ are similar, and write

$$\frac{\text{OC}}{1} = \frac{1}{\text{OA}} \qquad \text{or} \qquad \text{OC} = \frac{1}{\cos\theta}$$

**(f)** Use the fact that $\triangle\text{GOH}$ and $\triangle\text{DOB}$ are similar, and write

$$\frac{\text{GH}}{1} = \frac{1}{\text{DB}} \qquad \text{or} \qquad \text{GH} = \frac{1}{\tan\theta}$$

**49.** If we consider a triangle with opposite side of length 3 and hypotenuse 5, we can use the Pythagorean theorem to find the length of the adjacent side as

$$\sqrt{5^2 - 3^2} = 4.$$

This gives a triangle with sides 3 and 4, and hypotenuse 5. Since tangent is negative in the fourth quadrant, $\tan\theta = -\frac{3}{4}$. See Figure 8.24.

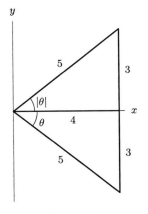

**Figure 8.24**

**53. (a)** The slope of the line is $m = \tan(5\pi/6) = -1/\sqrt{3}$, and the $y$-intercept is $b = 2$, so $y = (-1/\sqrt{3})x + 2$.
   **(b)** Solve $0 = (-1/\sqrt{3})x + 2$ to get $x = 2\sqrt{3}$.

**57. (a)** The period is $p = 80$, so $b = 2\pi/80$. The amplitude is $|a| = 300$, and the curve resembles an upside-down cosine, so $a = -300$. The midline is $k = 600$.

   Putting this information together gives $y = 600 - 300\cos(2\pi x/80)$.
   **(b)** These points fall on the line $y = 475$.

   The period is $p = 80$, so $b = 2\pi/80$. The amplitude is $|a| = 300$, and the curve resembles an upside-down cosine, so $a = -300$. The midline is $k = 600$. Thus, a formula for the sinusoidal function is $y = 600 - 300\cos(2\pi x/80)$, so we obtain one possible solution as follows:

$$600 - 300\cos\left(\frac{2\pi}{80}x\right) = 475$$

$$-300\cos\left(\frac{2\pi}{80}x\right) = 475 - 600 = -125$$

$$\cos\left(\frac{2\pi}{80}x\right) = \frac{-125}{-300} = \frac{5}{12}$$

$$\frac{2\pi}{80}x = \cos^{-1}\left(\frac{5}{12}\right)$$

$$x = \frac{80}{2\pi}\cos^{-1}\left(\frac{5}{12}\right)$$

$$= 14.5279.$$

We denote this solution by $x_0$, and it gives the $x$-coordinate of the left-most point. By symmetry, we see that

$$x_1 = 80 - x_0 = 65.4721$$
$$x_2 = 80 + x_0 = 94.5279.$$

**61.** First, by looking at the graph of $f$, we see that the difference between the maximum and minimum output values is $8 - (-2) = 10$, so the amplitude is 5. Therefore, the midline of the graph is given by the line $y = -2 + 5 = 3$, so we have $A = \pm 5$ and $k = 3$. We also note that the graph of $f$ completes 1.5 cycles between $x = -2$ and $x = 16$, so the period of the graph is 12, from which it follows that $B = 2\pi/12 = \pi/6$.

   Combining these observations, we see that we can take the four formulas to be horizontal translations of the functions $y_1 = 5\cos((\pi/6)x) + 3, y_2 = -5\cos((\pi/6)x) + 3, y_3 = 5\sin((\pi/6)x) + 3$, and $y_4 = -5\sin((\pi/6)x) + 3$, all of which have the same amplitude, period, and midline as $f$. After sketching $y_1$, we see that we can obtain the graph of $f$ by shifting the graph of $y_1$ 2 units to the left; therefore, $f_1(x) = 5\cos((\pi/6)(x + 2)) + 3$. Similarly, we can obtain the graph of $f$ by shifting the graph of $y_2$ to the right 4 units, so $f_2(x) = -5\cos((\pi/6)(x - 4)) + 3$, or by shifting the graph of $y_3$ to the right 7 units, so $f_3(x) = 5\sin((\pi/6)(x - 7)) + 3$. Finally, a sketch of $y_4$ reveals that we can obtain the graph of $f$ by shifting the graph of $y_4$ to the right 1 unit, so $f_4(x) = -5\sin((\pi/6)(x - 1)) + 3$. Answers may vary.

**65.** We first solve for $\sin\alpha$,

$$3\sin^2\alpha + 4 = 5$$
$$3\sin^2\alpha = 1$$
$$\sin^2\alpha = \frac{1}{3}$$
$$\sin\alpha = \pm\sqrt{\frac{1}{3}}$$

$$\sin\alpha = \sqrt{\frac{1}{3}} \qquad \sin\alpha = -\sqrt{\frac{1}{3}}$$
$$\alpha = 0.616, \ 2.526 \qquad \alpha = 3.757, \ 5.668$$

**69.** We can approximate this angle by using $s = r\theta$. The arc length is approximated by the moon diameter; and the radius is the distance to the moon. Therefore $\theta = s/r = 2160/238,860 \approx 0.009$ radians. Change this to degrees to get $\theta = 0.009(180/\pi) \approx 0.516°$. Note that we could also consider the radius to cut across the moon's center, in which case the radius would be $r = 238,860 + 2160/2 = 239,940$. The difference in the two answers is negligible.

**73.** The function has a maximum of 3000, a minimum of 1200 which means the upward shift is $\frac{3000+1200}{2} = 2100$. A period of eight years means the angular frequency is $\frac{\pi}{4}$. The amplitude is $|A| = 3000 - 2100 = 900$. Thus a function for the population would be an inverted cosine and $f(t) = -900\cos((\pi/4)t) + 2100$.

## CHECK YOUR UNDERSTANDING

**1.** False. One radian is about 57 degrees.

**5.** False. Use $s = r\theta$ to get, $s = 3 \cdot (\pi/3) = \pi$.

**9.** False. $\sin\frac{\pi}{6} = \frac{1}{2}$.

**13.** False. The amplitude is (maximum − minimum)/2 = $(6 - 2)/2 = 2$.

**17.** False. Amplitude is always positive. In this case it is equal to three.

**21.** True. The minimum $y$-value is when $\cos x = -1$. It is 15.

**25.** False, since $\sin(\pi x)$ has period $2\pi/\pi = 2$.

**29.** False. The period is $\frac{1}{3}$ that of $y = \cos x$.

**33.** True. The amplitude is $\frac{1}{2}$, the period is $\pi$ and the midline is $y = 1$ and its horizontal shift is correct.

**37.** True. Because the angles $\theta$ and $\theta + \pi$ determine the same line through the origin and hence have the same slope, which is the tangent.

**41.** True. We have $\sec\pi = 1/\cos\pi = 1/(-1) = -1$.

**45.** True. Since $y = \arctan(-1) = -\pi/4$, we have $\sin(-\pi/4) = -\sqrt{2}/2$.

**49.** True, because $\sin x$ is an odd function.

**53.** True. If $g(t) = f(t - \pi/2)$, the graph of $g$ is related to the graph of $f$ by a horizontal shift of $\pi/2$ units to the right.

**57.** True. If $\cos t = 1$, then $\sin t = 0$ so $\tan t = 0/1 = 0$.

**61.** True

**65.** False. The graph of $r = 1$ is the unit circle.

**69.** True. The point is on the $y$ axis three units down from the origin. Thus $r = 3$ and $\theta = 3\pi/2$. In polar coordinates this is $(3, 3\pi/2)$.

**73.** False, since $(1 + i)^2 = 2i$ is not real.

**77.** True. This is Euler's formula, fundamental in higher mathematics.

# CHAPTER NINE

## Solutions for Section 9.1

### Exercises

**1.** We have:

$$\tan t \cos t - \frac{\sin t}{\tan t} = \frac{\sin t}{\cos t} \cdot \cos t - \frac{\sin t}{\left(\dfrac{\sin t}{\cos t}\right)} \quad \text{because } \tan t = \frac{\sin t}{\cos t}$$

$$= \sin t - \sin t \cdot \frac{\cos t}{\sin t}$$

$$= \sin t - \cos t.$$

**5.** Writing $\sin 2\alpha = 2 \sin \alpha \cos \alpha$, we have

$$\frac{\sin 2\alpha}{\cos \alpha} = \frac{2 \sin \alpha \cos \alpha}{\cos \alpha} = 2 \sin \alpha.$$

**9.** Combining terms and using $\cos^2 \phi + \sin^2 \phi = 1$, we have

$$\frac{\cos \phi - 1}{\sin \phi} + \frac{\sin \phi}{\cos \phi + 1} = \frac{(\cos \phi - 1)(\cos \phi + 1) + \sin^2 \phi}{\sin \phi(\cos \phi + 1)} = \frac{\cos^2 \phi - 1 + \sin^2 \phi}{\sin \phi(\cos \phi + 1)} = \frac{0}{\sin \phi(\cos \phi + 1)} = 0$$

**13.** We have:

$$\frac{3 \sin(\phi + 1)}{4 \cos(\phi + 1)} = \frac{3}{4} \cdot \frac{\sin(\phi + 1)}{\cos(\phi + 1)}$$

$$= \frac{3}{4} \tan(\phi + 1).$$

**17.** The relevant identities are $\cos^2 \theta + \sin^2 \theta = 1$ and $\cos 2\theta = \cos^2 \theta - \sin^2 \theta = 2 \cos^2 \theta - 1 = 1 - 2 \sin^2 \theta$. See Table 9.1.

**Table 9.1**

| $\theta$ in rad. | $\sin^2 \theta$ | $\cos^2 \theta$ | $\sin 2\theta$ | $\cos 2\theta$ |
|---|---|---|---|---|
| 1 | 0.708 | 0.292 | 0.909 | $-0.416$ |
| $\pi/2$ | 1 | 0 | 0 | $-1$ |
| 2 | 0.827 | 0.173 | $-0.757$ | $-0.654$ |
| $5\pi/6$ | $1/4$ | $3/4$ | $-\sqrt{3}/2$ | $1/2$ |

### Problems

**21. (a)** $\cos 2\theta = 1 - 2 \sin^2 \theta = 1 - 2(1 - \cos^2 \theta) = 1 - 2 + 2 \cos^2 \theta = 2 \cos^2 \theta - 1 = 2(\cos \theta)^2 - 1.$
**(b)** $\cos 2\theta = 1 - 2 \sin^2 \theta = (1 - \sin^2 \theta) - \sin^2 \theta = \cos^2 \theta - \sin^2 \theta = (\cos \theta)^2 - (\sin \theta)^2.$

**25.** In order to get tan to appear, divide by $\cos x \cos y$:

$$\frac{\sin x \cos y + \cos x \sin y}{\cos x \cos y - \sin x \sin y} = \frac{\dfrac{\sin x \cos y}{\cos x \cos y} + \dfrac{\cos x \sin y}{\cos x \cos y}}{\dfrac{\cos x \cos y}{\cos x \cos y} - \dfrac{\sin x \sin y}{\cos x \cos y}} = \frac{\tan x + \tan y}{1 - \tan x \tan y}$$

**29.** Using the trigonometric identity $\tan(2\theta) = 2\tan\theta/(1 - \tan^2\theta)$, we have

$$
\begin{aligned}
\tan(2\theta) + \tan\theta &= 0 \\
\frac{2\tan\theta}{1 - \tan^2\theta} &= -\tan\theta \\
2\tan\theta &= -\tan\theta + \tan^3\theta \\
\tan^3\theta - 3\tan\theta &= 0 \\
\tan\theta(\tan^2\theta - 3) &= 0 \\
\tan\theta = 0 \quad\text{or}\quad \tan\theta &= \pm\sqrt{3}.
\end{aligned}
$$

If $\tan\theta = 0$, then we have three solutions: $\theta = 0$ and $\theta = \pi$ and $\theta = 2\pi$. On the other hand, if $\tan\theta = \sqrt{3}$, we first calculate the associated reference angle, which is $\tan^{-1}(\sqrt{3}) = \pi/3$. Using a graph of the tangent function on the interval $0 \le \theta \le 2\pi$, we see that the two solutions to $\tan\theta = \sqrt{3}$ are given by $\theta = \pi/3$ and $\theta = \pi + \pi/3 = 4\pi/3$. Finally, if $\tan\theta = -\sqrt{3}$, we again have a reference angle of $\pi/3$, and the two solutions to $\tan\theta = -\sqrt{3}$ are given by $\theta = \pi - \pi/3 = 2\pi/3$ and $\theta = 2\pi - \pi/3 = 5\pi/3$. Combining the above observations, we see that there are seven solutions to the original equation: $0, \pi, \pi/3, 4\pi/3, 2\pi/3, 5\pi/3$, and $2\pi$.

**33.** Not an identity. False for $x = 2$.

**37.** Identity. $\dfrac{\sin 2x}{1 + \cos 2x} = \dfrac{2\sin x \cos x}{1 + 2\cos^2 x - 1} = \dfrac{2\sin x \cos x}{2\cos^2 x} = \dfrac{\sin x}{\cos x} = \tan x.$

**41.** Identity. $\sin x \tan x = \sin x \cdot \dfrac{\sin x}{\cos x} = \dfrac{\sin^2 x}{\cos x} = \dfrac{1 - \cos^2 x}{\cos x}.$

**45.** If we let $\theta = 1$, then we have

$$
\frac{\sin\theta}{\cos\theta} - \frac{\cos\theta}{\sin\theta} = \frac{\sin 1}{\cos 1} - \frac{\cos 1}{\sin 1} = 0.915 \ne -0.458 = \frac{\cos 2}{\sin 2} = \frac{\cos(2\theta)}{\sin(2\theta)}.
$$

Therefore, since the equation is not true for $\theta = 1$, it is not an identity.

**49. (a)** By the Pythagorean theorem, the side adjacent to $\theta$ has length $\sqrt{1 - y^2}$. So

$$
\cos\theta = \sqrt{1 - y^2}/1 = \sqrt{1 - y^2}.
$$

**(b)** Since $\sin\theta = y/1$, we have

$$
\tan\theta = \frac{y}{\sqrt{1 - y^2}}.
$$

**(c)** Using the double angle formula,

$$
\cos(2\theta) = 1 - 2\sin^2\theta = 1 - 2y^2.
$$

**(d)** Supplementary angles have equal sines:

$$
\sin(\pi - \theta) = \sin\theta = y.
$$

**(e)** Since $\cos(\pi/2 - \theta) = y$, we have $\sin(\cos^{-1}(y)) = \sin(\pi/2 - \theta) = \sqrt{1 - y^2}$. So

$$
\sin^2(\cos^{-1}(y)) = 1 - y^2.
$$

**53. (a)** Let $\theta = \cos^{-1} x$, so $\cos\theta = x$. Then, since $0 \le \theta \le \pi$, $\sin\theta = \sqrt{1 - x^2}$ and $\tan\theta = \dfrac{\sqrt{1 - x^2}}{x}$, and $\tan(2\cos^{-1} x) = \tan 2\theta = \dfrac{2\tan\theta}{1 - \tan^2\theta}$. Now $1 - \tan^2\theta = 1 - \dfrac{1 - x^2}{x^2} = \dfrac{2x^2 - 1}{x^2}$, so $\tan 2\theta = \dfrac{2\sqrt{1 - x^2}}{x} \cdot \dfrac{x^2}{2x^2 - 1} = \dfrac{2x\sqrt{1 - x^2}}{2x^2 - 1}.$

**(b)** Let $\theta = \tan^{-1} x$, so $\tan\theta = x$. Then $\sin\theta = \dfrac{x}{\sqrt{1 + x^2}}$ and $\cos\theta = \dfrac{1}{\sqrt{1 + x^2}}$, so $\sin(2\tan^{-1} x) = \sin 2\theta = 2\sin\theta\cos\theta = 2\left(\dfrac{x}{\sqrt{1 + x^2}}\right)\left(\dfrac{1}{\sqrt{1 + x^2}}\right) = \dfrac{2x}{1 + x^2}.$

**57.** **(a)** Since $-\pi \le t < 0$ we have $0 < -t \le \pi$ so the double angle formula for sine can be used for the angle $\theta = -t$. Therefore $\sin 2\theta = 2\sin\theta\cos\theta$ tells us that

$$\sin(-2t) = 2\sin(-t)\cos(-t).$$

**(b)** Since sine is odd, we have $\sin(-2t) = -\sin 2t$. Since sine is odd and cosine is even, we have

$$2\sin(-t)\cos(-t) = 2(-\sin t)\cos t = -2\sin t\cos t.$$

Substitution of these results into the results of part (a) shows that

$$-\sin(2t) = -2\sin t\cos t.$$

Multiplication by $-1$ gives

$$\sin 2t = 2\sin t\cos t.$$

**61.** We know that $\cos(\pi/3) = \cos 60° = 1/2$ so $\cos^{-1}\left(\frac{1}{2}\right) = \pi/3$. This gives

$$\cos\left(\cos^{-1}\left(\frac{1}{2}\right)\right) = \cos\left(\frac{\pi}{3}\right) = \frac{1}{2}.$$

This is not at all surprising: after all, what we are saying is that the cosine of inverse cosine of a number is that number. However, the situation is not as straightforward as it may appear. For example, to evaluate the expression $\cos^{-1}\left(\cos\left(\frac{5\pi}{3}\right)\right)$, we write

$$\cos^{-1}\left(\cos\left(\frac{5\pi}{3}\right)\right) = \cos^{-1}\left(\frac{1}{2}\right),$$

because the cosine of $5\pi/3$ is $1/2$. And since again $\cos^{-1}\left(\frac{1}{2}\right) = \pi/3$, we have

$$\cos^{-1}\left(\cos\left(\frac{5\pi}{3}\right)\right) = \cos^{-1}\left(\frac{1}{2}\right) = \frac{\pi}{3}.$$

Thus, we see that the inverse cosine of the cosine of an angle does not necessarily equal that angle.

## Solutions for Section 9.2

### Exercises

**1.** We have $A = \sqrt{8^2 + (-6)^2} = \sqrt{100} = 10$. Since $\cos\phi = 8/10 = 0.8$ and $\sin\phi = -6/10 = -0.6$, we know that $\phi$ is in the fourth quadrant. Thus,

$$\tan\phi = -\frac{6}{8} = -0.75 \quad \text{and} \quad \phi = \tan^{-1}(-0.75) = -0.644,$$

so $8\sin t - 6\cos t = 10\sin(t - 0.644)$.

**5.** Write $\sin 15° = \sin(45° - 30°)$, and then apply the appropriate trigonometric identity.

$$\begin{aligned}
\sin 15° &= \sin(45° - 30°)\\
&= \sin 45°\cos 30° - \sin 30°\cos 45°\\
&= \frac{\sqrt{6}}{4} - \frac{\sqrt{2}}{4}
\end{aligned}$$

Similarly, $\sin 75° = \sin(45° + 30°)$.

$$\begin{aligned}
\sin 75° &= \sin(45° + 30°)\\
&= \sin 45°\cos 30° + \sin 30°\cos 45°\\
&= \frac{\sqrt{6}}{4} + \frac{\sqrt{2}}{4}
\end{aligned}$$

Also, note that $\cos 75° = \sin(90° - 75°) = \sin 15°$, and $\cos 15° = \sin(90° - 15°) = \sin 75°$.

**9.** Since $105°$ is $60° + 45°$, we can use the sum-of-angle formula for sine and say that

$$\sin 105 = \sin 60 \cos 45 + \sin 45 \cos 60 = (\sqrt{3}/2)(\sqrt{2}/2) + (\sqrt{2}/2)(1/2) = (\sqrt{6} + \sqrt{2})/4.$$

## Problems

**13. (a)** $\cos(t - \pi/2) = \cos t \cos \pi/2 + \sin t \sin \pi/2 = \cos t \cdot 0 + \sin t \cdot 1 = \sin t.$
**(b)** $\sin(t + \pi/2) = \sin t \cos \pi/2 + \sin \pi/2 \cos t = \sin t \cdot 0 + 1 \cdot \cos t = \cos t.$

**17.** For the sine, we have $\sin 2t = \sin(t + t) = \sin t \cos t + \sin t \cos t = 2 \sin t \cos t.$ This is the double angle formula for sine.

For cosine, we have $\cos 2t = \cos(t+t) = \cos t \cos t - \sin t \sin t = \cos^2 t - \sin^2 t.$ This is the double angle formula for cosine.

**21.** We manipulate the equation for the average rate of change as follows:

$$\frac{\cos(x + h) - \cos x}{h} = \frac{\cos x \cos h - \sin x \sin h - \cos x}{h}$$

$$= \frac{\cos x \cos h - \cos x}{h} - \frac{\sin x \sin h}{h}$$

$$= \cos x \left( \frac{\cos h - 1}{h} \right) - \sin x \left( \frac{\sin h}{h} \right).$$

**25. (a)** Since $\triangle CAD$ and $\triangle CDB$ are both right triangles, it is easy to calculate the sine and cosine of their angles:

$$\sin \theta = \frac{c_1}{b}$$

$$\cos \theta = \frac{h}{b}$$

$$\sin \phi = \frac{c_2}{a}$$

$$\cos \phi = \frac{h}{a}.$$

**(b)** We can calculate the areas of the triangles using the formula Area $=$ Base $\cdot$ Height:

$$\text{Area } \triangle CAD = \frac{1}{2} c_1 \cdot h$$

$$= \frac{1}{2} (b \sin \theta)(a \cos \phi),$$

$$\text{Area } \triangle CDB = \frac{1}{2} c_2 \cdot h$$

$$= \frac{1}{2} (a \sin \phi)(b \cos \theta).$$

**(c)** We find the area of the whole triangle by summing the area of the two constituent triangles:

$$\text{Area } \triangle ABC = \text{Area } \triangle CAD + \text{Area } CDB$$

$$= \frac{1}{2} (b \sin \theta)(a \cos \phi) + \frac{1}{2} (a \sin \phi)(b \cos \theta)$$

$$= \frac{1}{2} ab (\sin \theta \cos \phi + \sin \phi \cos \theta)$$

$$= \frac{1}{2} ab \sin(\theta + \phi)$$

$$= \frac{1}{2} ab \sin C.$$

# Solutions for Section 9.3

## Exercises

1.

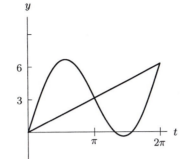

For the graphs to intersect, $t + 5\sin t = t$. So $\sin t = 0$, or $t =$ any integer multiple of $\pi$.

## Problems

5. **(a)** We start the time count on Jan 1, so substituting $t = 0$ into $f(t)$ gives us the value of $b$, since both $mt$ and $A\sin\dfrac{\pi t}{6}$ are equal to zero when $t = 0$. Thus, $b = f(0) = 20$. We see that in the 12 month period between Jan 1 and Jan 1 the value of the stock rose by \$30.00. Therefore, the linear component grows at the rate of \$30.00/year, or in terms of months, $30/12 = \$2.50$/month. So $m = 2.5$. Thus we have

$$P = f(t) = 2.5t + 20 + A\sin\frac{\pi t}{6}.$$

At an arbitrary data point, say $(Apr1, 37.50)$, we can solve for $A$. Since January 1 corresponds to $t = 0$, April 1 is $t = 3$. We have

$$37.50 = f(3) = 2.5(3) + 20 + A\sin\frac{3\pi}{6} = 7.5 + 20 + A\sin\frac{\pi}{2} = 27.5 + A.$$

Simplifying gives $A = 10$, and the function is

$$f(t) = 2.5t + 20 + 10\sin\frac{\pi t}{6}.$$

**(b)** The stock appreciates the most during the months when the sine function climbs the fastest. By looking at Figure 9.1 we see that this occurs roughly when $t = 0$ and $t = 11$, January and December.

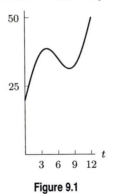

**Figure 9.1**

**(c)** Again, we look to Figure 9.1 to see when the graph actually decreases. It seems that the graph is decreasing roughly between the fourth and eighth months, that is, between May and September.

**9.** Use the sum-of-angles identity for cosine on the second term to get

$$I\cos(\omega_c + \omega_d)t = I\cos\omega_c t \cos\omega_d t - I\sin\omega_c t \sin\omega_d t$$

and factor the term $\cos\omega_c t$ to show the equality.

## Solutions for Chapter 9 Review

### Exercises

**1.** We have:

$$(1 - \sin t)(1 - \cos t) - \cos t \sin t = \underbrace{1 - \cos t - \sin t + \sin t \cos t}_{(1-\sin t)(1-\cos t)} - \cos t \sin t \quad \text{multiply out}$$
$$= 1 - \cos t - \sin t.$$

**5.** We have:

$$\frac{\sec t}{\csc t} = \frac{\left(\frac{1}{\cos t}\right)}{\left(\frac{1}{\sin t}\right)}$$
$$= \frac{1}{\cos t} \cdot \sin t$$
$$= \tan t.$$

**9.** Using $1 - \cos^2\theta = \sin^2\theta$, we have

$$\frac{1 - \cos^2\theta}{\sin\theta} = \frac{\sin^2\theta}{\sin\theta} = \sin\theta.$$

### Problems

**13.** They are both right. The first student meant that $\sin 2\theta = 2\sin\theta$ is not an identity meaning that it is not true for *all* $\theta$. The second student had found one value for $\theta$ for which it was true.

**17.** Since $\sin\theta = 1/\csc\theta$, we have $\sin\theta = 1/94$. Using the Pythagorean Identity, $\cos^2\theta + \sin^2\theta = 1$, we have

$$\cos^2\theta + \left(\frac{1}{94}\right)^2 = 1$$
$$\cos^2\theta = 1 - \frac{1}{94^2}$$
$$\cos\theta = \pm\sqrt{\frac{8835}{8836}} = \pm\frac{\sqrt{8835}}{94}.$$

Since $0 \le \theta \le \pi/2$, we know that $\cos\theta \ge 0$, so $\cos\theta = \sqrt{8835}/94$.

Using the identity $\tan\theta = \sin\theta/\cos\theta$, we see that $\tan\theta = (1/94)/(\sqrt{8835}/94) = 1/\sqrt{8835}$.

**21.** We will use one of the three double angle formulas for cosine to solve this equation algebraically. Since the equation involves $\sin\theta$, we will try the double angle formula for cosine which involves the sine:

$$\cos 2\theta = 1 - 2\sin^2\theta.$$

This gives

$$1 - 2\sin^2\theta = \sin\theta,$$

and we can rewrite this equation as follows:

$$2(\sin\theta)^2 + \sin\theta - 1 = 0.$$

Factoring, we have

$$(2\sin\theta - 1)(\sin\theta + 1) = 0.$$

The fact that the product $(2\sin\theta - 1)(\sin\theta + 1)$ equals zero implies that

$$2\sin\theta - 1 = 0 \quad \text{or} \quad \sin\theta + 1 = 0.$$

Solving $\sin\theta + 1 = 0$, we have $\sin\theta = -1$. This means $\theta = 3\pi/2$. Solving the other equation, we have

$$2\sin\theta - 1 = 0$$
$$\sin\theta = \frac{1}{2}.$$

This gives

$$\theta = \frac{\pi}{6} \quad \text{or} \quad \theta = \pi - \frac{\pi}{6} = \frac{5\pi}{6}.$$

In summary, there are three solutions for $0 \le \theta < 2\pi : \theta = \pi/6, 5\pi/6$, and $3\pi/2$. Figure 9.2 illustrates these solutions graphically as the points where the graphs of $\cos 2\theta$ and $\sin\theta$ intersect.

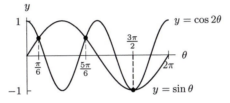

**Figure 9.2**: There are three solutions to the equation $\cos 2\theta = \sin\theta$, for $0 \le \theta < 2\pi$

**25.** Start with the left side, which is more complex, and reduce it using the Pythagorean identity:

$$(\sin^2 2\pi t + \cos^2 2\pi t)^3 = (1)^3 = 1.$$

**29.** We first solve for $\cos\alpha$,

$$3\cos^2\alpha + 2 = 3 - 2\cos\alpha$$
$$3\cos^2\alpha + 2\cos\alpha - 1 = 0$$
$$(3\cos\alpha - 1)(\cos\alpha + 1) = 0$$

$$3\cos\alpha - 1 = 0 \qquad\qquad \cos\alpha + 1 = 0$$
$$\cos\alpha = \tfrac{1}{3} \qquad\qquad\qquad \cos\alpha = -1$$
$$\alpha = 1.231,\ 5.052 \qquad\qquad \alpha = \pi$$

**33.** Since $0 < \ln x < \frac{\pi}{2}$ and $0 < \ln y < \frac{\pi}{2}$, the angles represented by $\ln x$ and $\ln y$ are in the first quadrant. This means that both their sine and cosine values will be positive. Since $\ln(xy) = \ln x + \ln y$, we can write

$$\sin(\ln(xy)) = \sin(\ln x + \ln y).$$

By the sum-of-angle formula we have

$$\sin(\ln x + \ln y) = \sin(\ln x)\cos(\ln y) + \cos(\ln x)\sin(\ln y).$$

Since cosine is positive, we have

$$\cos(\ln x) = \sqrt{1 - \sin^2(\ln x)} = \sqrt{1 - \left(\frac{1}{3}\right)^2} = \frac{\sqrt{8}}{3}$$

and

$$\cos(\ln y) = \sqrt{1 - \sin^2(\ln y)} = \sqrt{1 - \left(\frac{1}{5}\right)^2} = \frac{\sqrt{24}}{5}.$$

Thus,

$$\sin(\ln x + \ln y) = \sin(\ln x)\cos(\ln y) + \cos(\ln x)\sin(\ln y)$$
$$= \left(\frac{1}{3}\right)\left(\frac{\sqrt{24}}{5}\right) + \left(\frac{\sqrt{8}}{3}\right)\left(\frac{1}{5}\right)$$
$$= \frac{\sqrt{24} + \sqrt{8}}{15}$$
$$\approx 0.515.$$

## CHECK YOUR UNDERSTANDING

**1.** True.

**5.** True. This is an identity. Substitute using $\tan^2\theta = \sin^2\theta / \cos^2\theta$ and simplify to obtain $2 = 2\sin^2\theta + 2\cos^2\theta$. Divide by 2 to reach the Pythagorean Identity.

**9.** True. There are many ways to prove this identity. We use the identity $\cos 2\theta = \cos^2\theta - \sin^2\theta$ to substitute in the right side of the equation. This becomes $\frac{1}{2}(1 - (\cos^2\theta - \sin^2\theta))$. Now substitute using $1 - \cos^2\theta = \sin^2\theta$ (a form of the Pythagorean identity.) The right side then simplifies to $\sin^2\theta$ which is the left side.

**13.** True. Start with the sine sum-of-angle identity:

$$\sin(\theta + \phi) = \sin\theta\cos\phi + \sin\phi\cos\theta$$

and let $\phi = \pi/2$, so

$$\sin(\theta + \pi/2) = \sin\theta\cos(\pi/2) + \sin(\pi/2)\cos\theta.$$

Simplify to

$$\sin(\theta + \pi/2) = \sin\theta \cdot 0 + 1 \cdot \cos\theta = \cos\theta.$$

**17.** True. Since $A\cos(Bt) = A\sin(Bt + \pi/2) = A\sin(B(t + \pi/(2B)))$, the graph of $A\cos(Bt)$ is a shift of $A\sin(Bt)$ to the left by $\pi/(2B)$.

**21.** True. It is assumed that $a_1$ and $a_2$ are nonzero. The amplitude of the single sine function is $A = \sqrt{a_1^2 + a_2^2}$. Thus, $A$ is greater than either $a_1$ or $a_2$.

**25.** True. Hertz is a measure of cycles per second and so a single cycle will take 1/60th of a second.

**29.** True. Since $0 \leq \cos^{-1}(x) \leq \pi$, we have $0 \leq \sin(\cos^{-1} x) \leq 1$. Thus, $\cos^{-1}(\sin(\cos^{-1} x))$ is an angle $\theta$ whose cosine is between 0 and 1. In addition, we have $0 \leq \theta \leq \pi$, as this is part of the definition of $\cos^{-1} x$. Hence $0 \leq \theta \leq \pi/2$.

# CHAPTER TEN

## Solutions for Section 10.1

### Exercises

**1.** We have
$$f(g(x)) = 2^{x/(x+1)}.$$

**5.** We have
$$
\begin{aligned}
w(x) &= p(p(x)) \\
&= p(\underbrace{2x+1}_{\text{input for } p}) \qquad \text{because } p(x) = 2x+1 \\
&= 2(2x+1)+1 \qquad \text{because } p(\text{input}) = 2 \cdot \text{input} + 1 \\
&= 4x+3.
\end{aligned}
$$

**9.** Since $r(x) = \sqrt{3x}$, we substitute $\sqrt{3x}$ for $x$ in $f(x)$, giving us $f(r(x)) = 3(\sqrt{3x})^2$, which simplifies to $f(r(x)) = 9x$.

**13.** Since $g(x) = 9x - 2$, we substitute $9x - 2$ for $x$ in $m(x)$, giving us $m(g(x)) = 4(9x - 2)$, which simplifies to $m(g(x)) = 36x - 8$, which we then substitute for $x$ in $f(x)$, giving $f(m(g(x))) = 3(36x - 8)^2$, which simplifies to $f(m(g(x))) = 3888x^2 - 1728x + 192$.

**17.** The inside function is $f(x) = \cos 2x$.

### Problems

**21.** The function $R(Y(q))$ gives revenue as a function of the quantity of fertilizer.

**25.** $g(x) = \sqrt{x}$ and $h(x) = 1 + \sqrt{x}$

**29.** It is easiest to find values of $h$, because we can use the fact that $h(x) = g(f(x))$:
$$
\begin{aligned}
h(0) &= g(f(0)) \\
&= g(2) \qquad \text{because } f(0) = 2 \\
&= 3.
\end{aligned}
$$

Next, we will find values of $f$. To find $f(1)$, we know the output of $g(f(1))$ must be the same as $h(1)$, or 0. Since 0 is the output of $g$, we see from the table that its input must be 1. This means the value of $f(1)$ must be 1:
$$
\begin{aligned}
g(f(1)) &= h(1) = 0 \quad \text{because } h(1) = 0 \\
g(\underbrace{f(1)}_{1}) &= 0 \qquad\qquad \text{because } g(1) = 0 \\
\text{so} \quad f(1) &= 1.
\end{aligned}
$$

Likewise, for $f(2)$, we see that
$$g(f(2)) = h(2) = 2 \quad \text{because } h(2) = 2$$

$$g(\underbrace{f(2)}_{4}) = 2 \qquad \text{because } g(4) = 2$$

$$\text{so} \quad f(2) = 4.$$

Finally, we will find the values of $g$. To find $g(0)$, we know that the input of $g$ is 0, so the output of $f$ must be zero. This means that $x = 3$, because $f(3) = 0$. Thus, $g(0)$ is the same as $g(f(3))$, which equals $h(3)$ or 1:

$$g(0) = g(\underbrace{f(3)}_{0}) \quad \text{because } f(3) = 0$$

$$= h(3) \qquad \text{because } h(3) = g(f(3))$$

$$= 1.$$

Likewise, to find $g(3)$, we have

$$g(3) = g(\underbrace{f(4)}_{3}) \quad \text{because } f(4) = 3$$

$$= h(4) \qquad \text{because } g(f(4)) = h(4)$$

$$= 4.$$

See Table 10.1.

**Table 10.1**

| $x$ | $f(x)$ | $g(x)$ | $h(x)$ |
|---|---|---|---|
| 0 | 2 | 1 | 3 |
| 1 | 1 | 0 | 0 |
| 2 | 4 | 3 | 2 |
| 3 | 0 | 4 | 1 |
| 4 | 3 | 2 | 4 |

**33.** Since $k(f(x)) = e^{f(x)}$, we have $e^{f(x)} = x$. Taking the natural log of both sides, we obtain the formula $f(x) = \ln x$.

**37.** Since $f(x + h) = 2^{x+h}$,

$$\frac{f(x + h) - f(x)}{h} = \frac{2^{x+h} - 2^x}{h}.$$

**41. (a)** See Figures 10.1 and 10.2.

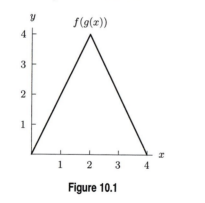

**Figure 10.1**

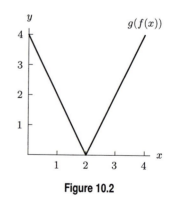

**Figure 10.2**

**(b)** From the graph, we see that $f(g(x))$ is increasing on the interval $0 < x < 2$.
**(c)** From the graph, we see that $g(f(x))$ is increasing on the interval $2 < x < 4$.

**45.** Reading values of the graph, we make an approximate table of values; we use these values to sketch Figure 10.3.

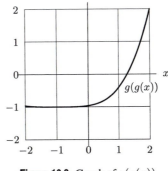

**Figure 10.3**: Graph of $g(g(x))$

| $x$ | $-2$ | $-1$ | 0 | 1 | 2 |
|---|---|---|---|---|---|
| $f(x)$ | $-2$ | $-0.3$ | $0.7$ | 1 | $0.7$ |
| $g(g(x))$ | $-1$ | $-1$ | $-1$ | $-0.4$ | 2 |

**49. (a)** Since $v(x) = x^2$ and $y$ can be written $\dfrac{1 + (x^2)}{2 + (x^2)}$, we take

$$u(x) = \frac{1 + x}{2 + x}.$$

**(b)** Since $v(x) = x^2 + 1$ and $y$ can be written $\dfrac{1 + x^2}{1 + 1 + x^2}$, we take

$$u(x) = \frac{x}{1 + x}.$$

**53. (a)** Since $u(x) = x^2$ and $y$ can be written as $y = (\sin x)^2$, we take $v(x) = \sin x$.
**(b)** Since $v(x) = x^2$ and $y$ can be written $y = \sin^2(\sqrt{x})^2$, we take $u(x) = \sin^2(\sqrt{x})$.

**57.** Substitute $q(p(t)) = \log(p(t))^2$ and solve:

$$
\begin{aligned}
\log(p(t))^2 &= 0 \\
\log(10(0.01)^t)^2 &= 0 \\
\log(10^2 \cdot (0.01)^{2t}) &= 0 \\
\log 10^2 + \log(0.01^{2t}) &= 0 \\
2 + 2t \log(0.01) &= 0 \\
2 + 2t(-2) &= 0 \\
2 - 4t &= 0 \\
t &= \frac{1}{2}
\end{aligned}
$$

**61.** We have

$$
\begin{aligned}
q(r(x)) &= \frac{8^{x^3}}{16^{x^2}} \\
&= 8^{x^3} \cdot 16^{-x^2} && \text{exponent rule} \\
&= (2^3)^{x^3} \cdot (2^4)^{-x^2} \\
&= 2^{3x^3} \cdot 2^{-4x^2} && \text{exponent rule} \\
&= 2^{3x^3 - 4x^2} && \text{exponent rule} \\
&= 2^{r(x)},
\end{aligned}
$$

so $q(x) = 2^x$.

**65.** Statements (a) and (e) must be true.

(a) To see why this statement must be true, suppose that $f$ is undefined at a particular value of $x$, say $x = 6$. Then $h(6) = g(f(6))$ is undefined, because $f(6)$ is undefined. Thus, in order for $h$ to be defined at all $x$, $f$ must be too.

(b) To see why this statement need not be true, suppose that $g(x) = \sqrt{x}$, so that $g$ is defined only for $x \geq 0$, and that $f(x) = x^2$. Then $h(x) = \sqrt{x^2}$. We see that $h$ is defined for all $x$, even though $g$ is not.

(c) To see why this statement need not be true, suppose that $f(x) = 3$, so that the graph of $f$ is a horizontal line, and similarly that $g(x) = 4$. Then $h(x) = g(f(x)) = g(3) = 4$. We see that even though the range of $f$ is the single value $y = 3$, the function $h$ can be defined for all $x$.

(d) From part (c), we see that if $f(x) = 3$ and $g(x) = 4$, we have $h(x) = 4$. This means that even if the range of $g$ is the single value $y = 4$, the function $h$ can be defined for all $x$.

(e) To see why this statement must be true, suppose there that $f(2) = 3$ but that $g(x) = 1/(x-3)$. We see that the range of $f$ includes 3, but that the domain of $g$ does not, because $g(3) = 1/0$. This means that $h$ is undefined at $x = 2$, because $h(2) = g(f(2)) = g(3)$, and $g(3)$ is undefined. The point is that: any legal output value of $f$ must be a legal input value of $g$, or otherwise $h$ will be undefined for some input values of $f$.

# Solutions for Section 10.2

## Exercises

**1.** It is not invertible.

**5.** It is not invertible.

**9.** One way to check that these functions are inverses is to make sure they satisfy the identities $g(g^{-1}(x)) = x$ and $g^{-1}(g(x)) = x$.

$$g(g^{-1}(x)) = 1 - \frac{1}{\left(1 + \dfrac{1}{1-x}\right) - 1}$$

$$= 1 - \frac{1}{\left(\dfrac{1}{1-x}\right)}$$

$$= 1 - (1 - x)$$

$$= x.$$

Also,

$$g^{-1}(g(x)) = 1 + \frac{1}{1 - \left(1 - \dfrac{1}{x-1}\right)}$$

$$= 1 + \frac{1}{\dfrac{1}{x-1}}$$

$$= 1 + x - 1 = x.$$

So the expression for $g^{-1}$ is correct.

**13.** Check using the two compositions

$$f(f^{-1}(x)) = e^{f^{-1}(x)/2} = e^{(2 \ln x)/2} = e^{\ln x} = x$$

and

$$f^{-1}(f(x)) = 2 \ln f(x) = 2 \ln e^{x/2} = 2(x/2) = x.$$

They are inverses of one another.

**17.** Let $y = \sqrt{x}$. Then $x = y^2$, and $h^{-1}(x) = x^2$.

**21.** Let $y = \sqrt{1 - 2x^2}$. We have

$$y = \sqrt{1 - 2x^2}$$
$$y^2 = 1 - 2x^2$$
$$2x^2 = 1 - y^2$$
$$x^2 = \frac{1 - y^2}{2}$$
$$x = \sqrt{\frac{1 - y^2}{2}}.$$

Thus, $l^{-1}(x) = \sqrt{\dfrac{1 - x^2}{2}}$.

**25.** Start with $x = j(j^{-1}(x))$ and substitute $y = j^{-1}(x)$. We have

$$x = j(y)$$
$$x = \sqrt{1 + \sqrt{y}}$$
$$x^2 = 1 + \sqrt{y}$$
$$x^2 - 1 = \sqrt{y}$$
$$(x^2 - 1)^2 = y$$

Therefore,

$$j^{-1}(x) = (x^2 - 1)^2.$$

**29.** Start with $x = h(h^{-1}(x))$ and substitute $y = h^{-1}(x)$. We have

$$x = h(y)$$
$$x = \log \frac{y + 5}{y - 4}$$
$$10^x = \frac{y + 5}{y - 4}$$
$$10^x(y - 4) = y + 5$$
$$10^x y - 4 \cdot 10^x = y + 5$$
$$10^x y - y = 5 + 4 \cdot 10^x$$
$$y(10^x - 1) = 5 + 4 \cdot 10^x$$
$$y = \frac{5 + 4 \cdot 10^x}{10^x - 1}$$
$$h^{-1}(x) = \frac{5 + 4 \cdot 10^x}{10^x - 1}.$$

## Problems

**33.** The inverse function $f^{-1}(P)$ gives the time, $t$, in years at which the population is $P$ thousand. Its units are years.

**37.** **(a)** $f(3) = 5^3 = 125$
$f^{-1}(\frac{1}{25}) = -2$ because $f(-2) = \frac{1}{25}$.

**(b)**

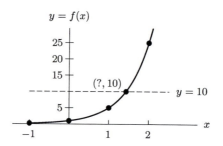

Using a calculator, $f^{-1}(10) \approx 1.43086$.

**41.** Solving for $t$ gives

$$P = 10e^{0.02t}$$
$$\frac{P}{10} = e^{0.02t}$$
$$0.02t = \ln\left(\frac{P}{10}\right)$$
$$t = \frac{\ln(P/10)}{0.02} = 50\ln(P/10).$$

The inverse function $t = f^{-1}(P) = 50\ln(P/10)$ gives the time, $t$, in years at which the population reaches $P$ million.

**45. (a)** The compositions are

$$f(g(x)) = f(\ln x) = e^{\ln x} = x \quad \text{and} \quad g(f(x)) = g(e^x) = \ln e^x = x,$$

which shows that the two functions are inverses of one another.

**(b)** The graph of the two functions is symmetric about the line $y = x$. See Figure 10.4.

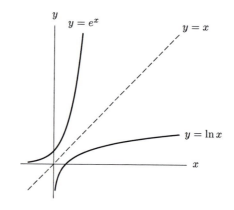

**Figure 10.4**

**49. (a)** Since $H(t)$ is a decreasing exponential function, we know that

$$H(t) = H_0 e^{-kt}$$

(Alternatively, we could have used $H(t) = H_0 a^t$, where $0 < a < 1$.) Since $H(0) = 200$, we have $H_0 = 200$. Thus, $H(t) = 200 e^{-kt}$. Since $H(2) = 20$, we have

$$200 e^{-2k} = 20$$
$$e^{-2k} = \frac{20}{200} = \frac{1}{10}$$
$$\ln e^{-2k} = \ln \frac{1}{10}$$
$$-2k = \ln \frac{1}{10}$$
$$k = -\frac{\ln \frac{1}{10}}{2} \approx 1.15129.$$

Thus,

$$H(t) = 200 e^{-1.15129t}.$$

**(b)** After one quarter hour, $t = 0.25$ and

$$H(0.25) = 200 e^{-1.15129(0.25)} = 149.979.$$

So the temperature dropped by about $200 - 149.979 = 50.021°$C in the first 15 minutes. After half an hour, $t = 0.5$ and

$$H(0.5) = 200 e^{-1.15129(0.5)} \approx 112.468.$$

So the temperature dropped by about $150 - 112.468 = 37.532°$C in the next 15 minutes.

**(c)** If $t = H^{-1}(y)$, then $t$ is the amount of time required for the brick's temperature to fall to $y°$C above room temperature. Letting $y = H(t)$, we have

$$y = 200 e^{-1.15129t}$$
$$e^{-1.15129t} = \frac{y}{200}$$
$$\ln e^{-1.15129t} = \ln \left( \frac{y}{200} \right)$$
$$-1.15129t = \ln \left( \frac{y}{200} \right)$$
$$t = \frac{\ln(y/200)}{-1.15129} \approx -0.86859 \ln \left( \frac{y}{200} \right) = H^{-1}(y).$$

**(d)** We need to evaluate $H^{-1}(5)$:

$$H^{-1}(5) = -0.86859 \ln \left( \frac{5}{200} \right) = 3.204 \text{ hours},$$

or about 3 hours and 12 minutes.

**(e)** The function $H(t)$ has a horizontal asymptote at $y = 0$ (the $t$-axis). The temperature of the brick approaches room temperature. Since $H(t)$ is the number of degrees above room temperature, $y = 0$ must be room temperature.

**53.** We have

$$f(-1) = -1 \cdot e^{-1} = -\frac{1}{e} \quad \text{so} \left( -1, -\frac{1}{e} \right) \text{ is on graph of } f$$
$$f(0) = 0 \cdot e^0 \quad = 0 \quad \text{so } (0, 0) \text{ is on graph of } f$$
$$f(1) = 1 \cdot e^1 \quad = e. \quad \text{so } (1, e) \text{ is on graph of } f$$

If $(x, y)$ is a point on the graph of $f$, then $(y, x)$ is a point on the graph of its inverse. This means three points on the graph of $W$ are $(-1/e, -1), (0, 0)$, and $(e, 1)$.

**57. (a)** $C(0)$ is the concentration of alcohol in the 100 ml solution after 0 ml of alcohol is removed. Thus, $C(0) = 99\%$.
   **(b)** Note that there are initially 99 ml of alcohol and 1 ml of water.

$$C(x) = \frac{\text{Concentration of alcohol}}{\text{after removing } x \text{ ml}} = \frac{\text{Amount of alcohol remaining}}{\text{Amount of solution remaining}}$$

$$= \frac{\begin{array}{c}\text{Original amount} \\ \text{of alcohol}\end{array} - \begin{array}{c}\text{Amount of} \\ \text{alcohol removed}\end{array}}{\begin{array}{c}\text{Original amount} \\ \text{of solution}\end{array} - \begin{array}{c}\text{Amount of alcohol} \\ \text{removed}\end{array}} = \frac{99 - x}{100 - x}.$$

   **(c)** If $y = C(x)$, then $x = C^{-1}(y)$. We have

$$y = \frac{99 - x}{100 - x}$$
$$y(100 - x) = 99 - x$$
$$100y - xy = 99 - x$$
$$x - xy = 99 - 100y$$
$$x(1 - y) = 99 - 100y$$
$$x = \frac{99 - 100y}{1 - y}.$$

   Thus, $C^{-1}(y) = \dfrac{99 - 100y}{1 - y}$.
   **(d)** The function $C^{-1}(y)$ tells us how much alcohol we need to remove in order to obtain a solution whose concentration is $y$.

**61.** Since $f$ is assumed to be an increasing function, its inverse is well-defined. This is an amount of caffeine. Notice that $r_c = f(q_c)$, and so $q_c = f^{-1}(r_c)$. Thus, $2f^{-1}(r_c) + 20 = 2q_c + 20$, making this 20 mg more caffeine than in 2 servings of coffee.

# Solutions for Section 10.3

## Exercises

**1. (a)** We have $f(x) + g(x) = x + 1 + 3x^2 = 3x^2 + x + 1$.
   **(b)** We have $f(x) - g(x) = x + 1 - 3x^2 = -3x^2 + x + 1$.
   **(c)** We have $f(x)g(x) = (x + 1)(3x^2) = 3x^3 + 3x^2$.
   **(d)** We have $f(x)/g(x) = (x + 1)/(3x^2)$.

**5. (a)** We have $f(x) + g(x) = x^3 + x^2$.
   **(b)** We have $f(x) - g(x) = x^3 - x^2$.
   **(c)** We have $f(x)g(x) = (x^3)(x^2) = x^5$.
   **(d)** We have $f(x)/g(x) = (x^3)/(x^2) = x$.

**9.** $h(x) = 2(2x - 1) - 3(1 - x) = 4x - 2 - 3 + 3x = 7x - 5$.

**13.** We have $f(x) = e^x(2x + 1) = 2xe^x + e^x$.

**17.** $f(x) + g(x) = \sin x + x^2$.

**21.** $g(f(x)) = g(\sin x) = \sin^2 x$.

## Problems

**25. (a)** Since the population consists only of men and women, the population size at any given time $t$ will be the sum of the numbers of women and the number of men at that particular time. Thus

$$p(t) = f(t) + g(t).$$

**(b)** In any given year the total amount of money that women in Canada earn is equal to the average amount of money one woman makes in that year times the number of women. Thus

$$m(t) = g(t) \cdot h(t).$$

**29.** To compute this table, note that since $f(x) = r(x) + t(x)$, then $f(-2) = r(-2) + t(-2) = 4 + 8 = 12$ and since $g(x) = 4 - 2s(x)$, then $g(-2) = 4 - 2(-2) = 4 + 4 = 8$. Repeat this process for each entry in the table.

**Table 10.2**

| $x$ | $f(x)$ | $g(x)$ | $h(x)$ | $j(x)$ | $k(x)$ | $l(x)$ |
|---|---|---|---|---|---|---|
| $-2$ | 12 | 8 | 32 | 2 | 16 | $-12$ |
| $-1$ | 10 | 0 | 25 | 0 | 25 | 15 |
| 0 | 13 | 8 | 42 | 0.5 | 36 | $-8$ |
| 1 | 4 | 0 | $-21$ | 5 | 49 | 1 |
| 2 | 10 | 8 | 16 | $-3$ | 64 | 4 |
| 3 | 22 | 0 | 117 | $-2$ | 81 | 35 |

**33.** We have:

$$q(0) = w^{-1}(v(0)) = w^{-1}(4) = 3 \quad \text{because } v(0) = 4 \text{ and } w(3) = 4$$
$$q(1) = w^{-1}(v(1)) = w^{-1}(3) = 2 \quad \text{because } v(1) = 3 \text{ and } w(2) = 3$$
$$q(2) = w^{-1}(v(2)) = w^{-1}(3) = 2 \quad \text{because } v(2) = 3 \text{ and } w(2) = 3$$
$$q(3) = w^{-1}(v(3)) = w^{-1}(5) = 5 \quad \text{because } v(3) = 5 \text{ and } w(5) = 5$$
$$q(4) = w^{-1}(v(4)) = w^{-1}(4) = 3 \quad \text{because } v(4) = 4 \text{ and } w(3) = 4$$
$$q(5) = w^{-1}(v(5)) = w^{-1}(4) = 3. \quad \text{because } v(5) = 4 \text{ and } w(3) = 4$$

See Table 10.3.

**Table 10.3**

| $x$ | 0 | 1 | 2 | 3 | 4 | 5 |
|---|---|---|---|---|---|---|
| $q(x)$ | 3 | 2 | 2 | 5 | 3 | 3 |

**37.** Since $f(a) = g(a)$, $h(a) = g(a) - f(a) = 0$. Similarly, $h(c) = 0$. On the interval $a < x < b$, $g(x) > f(x)$, so $h(x) = g(x) - f(x) > 0$. As $x$ increases from $a$ to $b$, the difference between $g(x)$ and $f(x)$ gets greater, becoming its greatest at $x = b$, then gets smaller until the difference is 0 at $x = c$. When $x < a$ or $x > b$, $g(x) < f(x)$ so $g(x) - f(x) < 0$. Subtract the length $e$ from the length $d$ to get the $y$-intercept. See Figure 10.5.

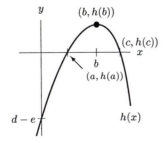

**Figure 10.5**

**41.** We have

$$H(x) = \frac{F(x)}{G(x)}$$

$$= \frac{e^{-x^2}}{x^4} = x^{-4} \cdot e^{-x^2}$$

$$h(x) = \frac{f(x)G(x) - G(x)g(x)}{(G(x))^2} .$$

$$= \frac{-2xe^{-x^2} \cdot x^4 - e^{-x^2} \cdot 4x^3}{(x^4)^2}$$

$$= \frac{-2x^5 e^{-x^2} - 4x^3 e^{-x^2}}{x^8} = -2x^{-3}\left(1 + 2x^{-2}\right)e^{-x^2} .$$

**45.** In order to evaluate $h(3)$, we need to express the formula for $h(x)$ in terms of $f(x)$ and $g(x)$. Factoring gives

$$h(x) = C^{2x}(kx^2 + B + 1).$$

Since $g(x) = C^{2x}$ and $f(x) = kx^2 + B$, we can re-write the formula for $h(x)$ as

$$h(x) = g(x) \cdot (f(x) + 1).$$

Thus,

$$h(3) = g(3) \cdot (f(3) + 1)$$
$$= 5(7 + 1)$$
$$= 40.$$

**49.** $g(q) < g(p)$ because the cost per square foot of building office space decreases as the total square footage increases. $g(p) < f(p)$ since the total cost of building more than one square foot is greater than the cost per square foot. $f(p) < f(q)$ since the total cost of building office space increases as the square footage increases. So

$$g(q) < g(p) < f(p) < f(q).$$

## Solutions for Chapter 10 Review

### Exercises

**1.** Using substitution we have $h(k(x)) = 2^{k(x)} = 2^{x^2}$ and $k(h(x)) = (h(x))^2 = (2^x)^2 = 2^{2x} = 4^x$.

**5.** Substituting the expression $x^2 + 1$ for the $x$ term in the formula for $h(x)$ gives $\sqrt{x^2 + 1}$.

**9.** **(a)** A graph of this function on a window which contains both positive and negative values of $x$ reveals that it fails the horizontal line test. Therefore, this function is not invertible.

**(b)** A graph reveals that the output values of this function oscillate back and forth between $y = -1$ and $y = 1$. It is therefore not invertible.

**(c)** A graph reveals that this function is always increasing. It passes the horizontal line test and is therefore invertible.

**13.** Start with $x = g(g^{-1}(x))$ and substitute $y = g^{-1}(x)$. We have

$$x = g(y)$$
$$x = e^{3y+1}$$
$$\ln x = \ln e^{(3y+1)}$$
$$\ln x = 3y + 1$$
$$\ln x - 1 = 3y$$
$$y = \frac{1}{3}(\ln x - 1).$$

Thus, $y = g^{-1}(x) = \frac{1}{3}(\ln x - 1)$.

**17.** Solving $y = g(x)$ for $x$ gives

$$y = \frac{x-2}{2x+3}$$
$$(2x+3)y = x - 2$$
$$2xy + 3y = x - 2$$
$$3y + 2 = x - 2xy$$
$$x(1 - 2y) = 3y + 2$$
$$x = \frac{3y+2}{1-2y},$$

so $g^{-1}(x) = \dfrac{3x+2}{1-2x}$.

**21.** Start with $x = s(s^{-1}(x))$ and substitute $y = s^{-1}(x)$. We have

$$x = s(y)$$
$$x = \frac{3}{2 + \log y}$$
$$\frac{x}{3} = \frac{1}{2 + \log y}$$
$$\frac{3}{x} = 2 + \log y$$
$$\frac{3}{x} - 2 = \log y$$
$$10^{\frac{3}{x}-2} = 10^{\log y}$$
$$10^{\frac{3}{x}-2} = y$$

So $s^{-1}(x) = 10^{(3/x)-2}$.

**25.** This function is not invertible. It does not pass the horizontal line test.

**29.** Since $h(x) = \sqrt{1-x}/x$ and $h^{-1}(x) = 1/(x^2 + 1)$, we have

$$h^{-1}(h(x)) = h^{-1}\left(\sqrt{\frac{1-x}{x}}\right)$$
$$= \frac{1}{\left(\sqrt{\dfrac{1-x}{x}}\right)^2 + 1}$$
$$= \frac{1}{\dfrac{1-x}{x} + 1}$$

$$= \frac{1}{\dfrac{1-x}{x} + \dfrac{x}{x}}$$

$$= \frac{1}{\dfrac{1-x+x}{x}} = \frac{1}{\left(\dfrac{1}{x}\right)} = x.$$

**33.** $g(g(x)) = g(2x - 1) = 2(2x - 1) - 1 = 4x - 3$

**37. (a)** $f(2x) = (2x)^2 + (2x) = 4x^2 + 2x$

**(b)** $g(x^2) = 2x^2 - 3$

**(c)** $h(1 - x) = \dfrac{(1-x)}{1 - (1-x)} = \dfrac{1-x}{x}$

**(d)** $(f(x))^2 = (x^2 + x)^2$

**(e)** Since $g(g^{-1}(x)) = x$, we have

$$2g^{-1}(x) - 3 = x$$
$$2g^{-1}(x) = x + 3$$
$$g^{-1}(x) = \frac{x+3}{2}.$$

**(f)** $(h(x))^{-1} = \left(\dfrac{x}{1-x}\right)^{-1} = \dfrac{1-x}{x}$

**(g)** $f(x)g(x) = (x^2 + x)(2x - 3)$

**(h)** $h(f(x)) = h(x^2 + x) = \dfrac{x^2 + x}{1 - (x^2 + x)} = \dfrac{x^2 + x}{1 - x^2 - x}$

**41.** To find $f(x)$, we add $m(x)$ and $n(x)$ and simplify: $m(x) + n(x) = 3x^2 - x + 2x = 3x^2 + x = f(x)$.

**45.** To find $j(x)$, we divide $m(x)$ by $n(x)$ and simplify: $(m(x))/n(x) = (3x^2 - x)/(2x) = 3x/2 - 1/2 = j(x)$.

**49.** Evaluate the two parts of the subtraction

$$h(g(x)) = \tan\left(2\left(\frac{(3x-1)^2}{4}\right)\right) = \tan\frac{(3x-1)^2}{2} \quad \text{and} \quad f(9x) = (9x)^{3/2} = 9^{3/2} \cdot x^{3/2} = 27x^{3/2}$$

and subtract

$$h(g(x)) - f(9x) = \tan\left(\frac{(3x-1)^2}{2}\right) - 27x^{3/2}.$$

## Problems

**53.** If $f(x) = u(v(x))$, then one solution is $u(x) = \sqrt{x}$ and $v(x) = 3 - 5x$.

**57.** One possible solution is $F(x) = u(v(x))$ where $u(x) = x^3$ and $v(x) = 2x + 5$.

**61.** The troughs (where the graph is below the $x$-axis) are reflected about the horizontal axis to become humps. The humps (where the graph is above the $x$-axis) are unchanged.

**65.** First, we have $h(0) = f(g(0)) = f(1) = 0$, which completes the first row of the table. From the information in the second row of the table, we see that $h(1) = 1$. Therefore, since $h(1) = f(g(1))$, we conclude that $f(g(1)) = 1$, which is equivalent to $f(x) = 1$ if we let $x = g(1)$. Since $f$ is invertible, our table indicates that $x = 2$ is the only solution to $f(x) = 1$. Therefore, $g(1) = x = 2$, which fills in the blank in the second row of the table. Finally, we have $h(2) = f(g(2)) = f(0) = 9$, which fills in the final entry in the table. See Table 10.4

**Table 10.4**

| $x$ | $f(x)$ | $g(x)$ | $h(x)$ |
|---|---|---|---|
| 0 | 9 | 1 | 0 |
| 1 | 0 | 2 | 1 |
| 2 | 1 | 0 | 9 |

**69.** The inverse function $g^{-1}(t)$ represents the velocity needed for a trip of $t$ hours. Its units are mph.

**73.** We take logarithms to help solve when $x$ is in the exponent:

$$2^{x+5} = 3$$
$$\ln(2^{x+5}) = \ln 3$$
$$(x+5)\ln 2 = \ln 3$$
$$x = \frac{\ln 3}{\ln 2} - 5.$$

**77.** Squaring eliminates square roots:

$$\sqrt{x + \sqrt{x}} = 3$$
$$x + \sqrt{x} = 9$$
$$\sqrt{x} = 9 - x \quad (\text{so} \quad x \le 9)$$
$$x = (9-x)^2 = 81 - 18x + x^2.$$

So $x^2 - 19x + 81 = 0$. The quadratic formula gives the solutions

$$x = \frac{19 \pm \sqrt{37}}{2}.$$

The only solution is $x = \dfrac{19 - \sqrt{37}}{2}$. The other solution is too large to satisfy the original equation.

**81. (a)** $f(g(a)) = f(a) = a$
**(b)** $g(f(c)) = g(c) = b$
**(c)** $f^{-1}(b) - g^{-1}(b) = 0 - c = -c$
**(d)** $0 < x \le a$

**85.**

$$p(x) = r(x)$$
$$2x - 3 = \frac{2x-1}{2x+1}$$
$$(2x-3)(2x+1) = 2x - 1$$
$$4x^2 - 4x - 3 = 2x - 1$$
$$4x^2 - 6x - 2 = 0$$
$$2x^2 - 3x - 1 = 0.$$

Using the quadratic formula, we have:

$$x = \frac{-(-3) \pm \sqrt{(-3)^2 - 4 \cdot 2 (-1)}}{2 \cdot 2}$$
$$= \frac{3 \pm \sqrt{17}}{4}.$$

**89. (a)** The function $f(x) = \sin^2 x$ is equal to $(u(x))^2$ but is not equal to $u(u(x))$. As an illustration of this, note that $f(\pi/2) = (\sin(\pi/2))^2 = 1$, but $u(u(\pi/2)) = \sin 1 \approx 0.84$. Since $f(1) \neq u(u(1))$, the functions $f(x)$ and $u(u(x))$ are not the same.

**(b)** First, we note that in the expression $p(x) = \sin(\cos^2 x)$, we are taking the composition of the sine function with $\cos^2 x$; we are not multiplying $\sin x$ by $\cos^2 x$. This tells us immediately that $u(x)(v(x))^2$ and $u(x)w(v(x))$ are not equal to $p(x)$. On the other hand, since $(v(x))^2 = w(v(x)) = \cos^2 x$, we see that $u((v(x))^2)$ and $u(w(v(x)))$ both equal $p(x)$, meaning that (ii) and (iii) are the only correct answers.

**(c)  (i)** We have $(u(x) + v(x))^2 = \sin^2 x + 2 \sin x \cos x + \cos^2 x$. Since $\sin^2 x + \cos^2 x = 1$ and $2 \sin x \cos x = \sin 2x$, our answer simplifies to $1 + \sin 2x$.

**(ii)** We have $(u(x))^2 + (v(x))^2 = \sin^2 x + \cos^2 x = 1$.

**(iii)** We have $u(x^2) + v(x^2) = \cos(x^2) + \sin(x^2)$, which cannot be simplified.

**93.** See Figure 10.6.

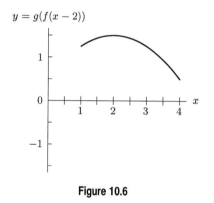

$y = g(f(x-2))$

**Figure 10.6**

**97. (a)** Since $f(x)$ is a linear function, its formula can be written in the form $f(x) = mx + b$, where $m$ represents the slope and $b$ represents the $y$-intercept. According to the graph, the $y$-intercept is 4. Since $(-2, 0)$ and $(0, 4)$ both lie on the line, we know that

$$m = \frac{y_2 - y_1}{x_2 - x_1} = \frac{4 - 0}{0 - (-2)} = \frac{4}{2} = 2.$$

So we know that the formula is $f(x) = 2x + 4$. Similarly, we can find the slope of $g(x)$, $\frac{0 - (-1)}{3 - 0} = \frac{1}{3}$, and the $y$-intercept, $-1$, so its formula is $g(x) = \frac{1}{3}x - 1$.

**(b)** To graph $h(x) = f(x) \cdot g(x)$, we first take note of where $f(x) = 0$ and $g(x) = 0$. At those places, $h(x) = 0$. Since the zero of $f(x)$ is $-2$ and the zero of $g(x)$ is 3, the zeros of $h(x)$ are $-2$ and 3. When $x < -2$, both $f(x)$ and $g(x)$ are negative, so we know that $h(x)$, their product, is positive. Similarly, when $x > 3$, both $f(x)$ and $g(x)$ are positive so $h(x)$ is positive. When $-2 < x < 3$, $f(x)$ is positive and $g(x)$ is negative, so $h(x)$ is negative. Also, since $h(x)$ is the product of two linear functions, we know that it is a quadratic function $h(x) = (2x+4)(\frac{1}{3}x-1) = \frac{2}{3}x^2 - \frac{2}{3}x - 4$. Putting these pieces of information together, we know that the graph of $h(x)$ is a parabola with zeros at $-2$ and 3 (and, therefore, an axis of symmetry at $x = \frac{1}{2}$) and that it is positive when $x < -2$ or $x > 3$ and negative when $-2 < x < 3$. [Note: since you know the axis of symmetry is $x = \frac{1}{2}$, you know that the $x$-coordinate of the vertex is $\frac{1}{2}$. You could find the $y$-coordinates of its vertex by finding $h(\frac{1}{2}) = (2(\frac{1}{2}) + 4)(\frac{1}{3}(\frac{1}{2}) - 1) = (1 + 4)(\frac{1}{6} - 1) = 5(-\frac{5}{6}) = -\frac{25}{6} = -4\frac{1}{6}$.] See Figure 10.7.

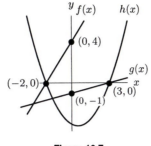

**Figure 10.7**

**101.** The statement is false. For example, if $f(x) = x$ and $g(x) = x^2$, then $f(x) \cdot g(x) = x^3$. In this case, $f(x) \cdot g(x)$ is an odd function, but $g(x)$ is an even function.

**105.** **(a)** One possible answer is:

$$y = \underbrace{6x}_{g(x)} \cdot \underbrace{e^{\overbrace{3x^2}^{G(x)}}}_{f(G(x))}$$

so  $f(x) = e^x$

$g(x) = 6x$

$G(x) = 3x^2.$

**(b)** One possible answer is:

$$y = -\frac{\sin\left(\sqrt{x}\right)}{2\sqrt{x}} = \underbrace{-\frac{1}{2\sqrt{x}}}_{g(x)} \cdot \underbrace{\sin\left(\overbrace{\sqrt{x}}^{G(x)}\right)}_{f(G(x))}$$

so  $f(x) = \sin x$

$g(x) = -\dfrac{1}{2\sqrt{x}}$

$G(x) = \sqrt{x}.$

**109.** This is an increasing function, because as $x$ increases, $f(x)$ increases, and as $f(x)$ increases, $f(f(x))$ increases.

**113.** **(a)** $f(8) = 2$, because 8 divided by 3 equals 2 with a remainder of 2. Similarly, $f(17) = 2$, $f(29) = 2$, and $f(99) = 0$.
  **(b)** $f(3x) = 0$ because, no matter what $x$ is, $3x$ will be divisible by 3.
  **(c)** No. Knowing, for example, that $f(x) = 0$ tells us that $x$ is evenly divisible by 3, but gives us no other information regarding $x$.
  **(d)** $f(f(x)) = f(x)$, because $f(x)$ equals either 0, 1, or 2, and $f(0) = 0$, $f(1) = 1$, and $f(2) = 2$.
  **(e)** No. For example, $f(1) + f(2) = 1 + 2 = 3$, but $f(1 + 2) = f(3) = 0$.

## CHECK YOUR UNDERSTANDING

**1.** False, since $f(4) + g(4) = \frac{1}{4} + \sqrt{4}$ but $(f + g)(8) = \frac{1}{8} + \sqrt{8}$.

**5.** True, since $g(f(x)) = g\left(\dfrac{1}{x}\right) = \sqrt{\dfrac{1}{x}}.$

**9.** True. Evaluate $\dfrac{f(3) + g(3)}{h(3)} = \dfrac{\frac{1}{3} + \sqrt{3}}{3 - 5}$ and simplify.

**13.** False. As a counterexample, let $f(x) = x^2$ and $g(x) = x + 1$. Then $f(g(x)) = (x + 1)^2 = x^2 + 2x + 1$, but $g(f(x)) = x^2 + 1$.

**17.** False. $f(x + h) = \dfrac{1}{x + h} \neq \dfrac{1}{x} + \dfrac{1}{h}$.

**21.** False. If $f(x) = ax^2 + bx + c$ and $g(x) = px^2 + qx + r$, then

$$f(g(x)) = f(px^2 + qx + r) = a(px^2 + qx + r)^2 + b(px^2 + qx + r) + c.$$

Expanding shows that $f(g(x))$ has an $x^4$ term.

**25.** True, since $g(f(2)) = g(1) = 3$ and $f(g(3)) = f(1) = 3$.

**29.** True. The function $g$ is not invertible if two different points in the domain have the same function value.

**33.** True. Each $x$ value has only one $y$ value.

**37.** True. The inverse of a function reverses the action of the function and returns the original value of the independent variable $x$.

# CHAPTER ELEVEN

## Solutions for Section 11.1

### Skill Refresher

**S1.** $\sqrt{36t^2} = (36t^2)^{1/2} = 36^{1/2} \cdot (t^2)^{1/2} = 6|t^1| = 6|t|$

**S5.** We have

$$10x^{5-2} = 2$$
$$10x^3 = 2$$
$$x^3 = 0.2$$
$$x = (0.2)^{1/3} = 0.585.$$

**S9.** False

### Exercises

1. Yes. Writing the function as

$$g(x) = \frac{(-x^3)^3}{6} = \frac{(-1)^3(x^3)^3}{6} = \frac{-x^9}{6} = -\frac{1}{6}x^9,$$

   we have $k = -1/6$ and $p = 9$.

5. No. This function cannot be written in the form $kx^p$ because the variable is in the exponent.

9. Since the graph is symmetric about the $y$-axis, the power function is even.

13. We use the form $y = kx^p$ and solve for $k$ and $p$. Using the point $(1, 3)$, we have $3 = k1^p$. Since $1^p$ is 1 for any $p$, we know that $k = 3$. Using our other point, we see that

$$13 = 3 \cdot 4^p$$
$$\frac{13}{3} = 4^p$$
$$\ln\left(\frac{13}{3}\right) = p\ln 4$$
$$\frac{\ln(13/3)}{\ln 4} = p$$
$$1.058 \approx p.$$

    So $y = 3x^{1.058}$.

17. Substituting into the general formula $c = kd^2$, we have $45 = k(3)^2$ or $k = 45/9 = 5$. So the formula for $c$ is

$$c = 5d^2.$$

    When $d = 5$, we get $c = 5(5)^2 = 125$.

21. We need to solve $f(x) = kx^p$ for $p$ and $k$. To solve for $p$, take the ratio of any two values of $f(x)$, say $f(3)$ over $f(2)$:

$$\frac{f(3)}{f(2)} = \frac{27}{12} = \frac{9}{4}.$$

Since $f(3) = k \cdot 3^p$ and $f(2) = k \cdot 2^p$, we have

$$\frac{f(3)}{f(2)} = \frac{k \cdot 3^p}{k \cdot 2^p} = \frac{3^p}{2^p} = \left(\frac{3}{2}\right)^p = \frac{9}{4}.$$

Since $\left(\frac{3}{2}\right)^p = \frac{9}{4}$, we know $p = 2$. Thus, $f(x) = kx^2$. To solve for $k$, use any point from the table. Note that $f(2) = k \cdot 2^2 = 4k = 12$, so $k = 3$. Thus, $f(x) = 3 \cdot x^2$.

25. (a) $\displaystyle\lim_{x \to \infty} x^{-4} = \lim_{x \to \infty} (1/x^4) = 0.$

    (b) $\displaystyle\lim_{x \to -\infty} 2x^{-1} = \lim_{x \to -\infty} (2/x) = 0.$

## Problems

29. (a) As $x \longrightarrow 0$ from the right, $x^{-3} \longrightarrow +\infty$, and $x^{1/3} \longrightarrow 0$.

    (b) As $x \longrightarrow \infty$, $x^{-3} \longrightarrow 0$, and $x^{1/3} \longrightarrow \infty$.

33. All four graphs contain the point $(1, 1)$, so they all four have $k = 1$. Judging from their graphs, $f$ and $g$ appear to have positive values of $p$. The graph of $g$ climbs faster as $x \to \infty$ so its value of $p$ is larger. The graphs of $v$ and $w$ have asymptotes, suggesting their values of $p$ are negative. The graph of $v$ approaches its horizontal asymptote faster than the graph of $w$, so its value of $p$ is "more negative."

    Therefore, ranking these functions in order of their power, $p$, gives: $v, w, f, g$.

37. (a) Since the cost of the fabric, $C(x)$, is directly proportional to the amount purchased, $x$, we know that the formula will be of the form

$$C(x) = kx.$$

    (b) Since 3 yards cost \$28.50, we know that $C(3) = \$28.50$. Thus, we have

$$28.50 = 3k$$
$$k = 9.5$$

    Our formula for the cost of $x$ yards of fabric is

$$C(x) = 9.5x.$$

    (c) Notice that the graph in Figure 11.1 goes through the origin.

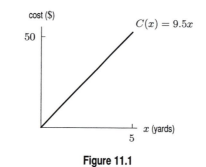

**Figure 11.1**

    (d) To find the cost of 5.5 yards of fabric, we evaluate $C(x)$ for $x = 5.5$:

$$C(5.5) = 9.5(5.5) = \$52.25.$$

41. Since the pitch is inversely proportional to the square root of the density, we have

$$P = k\frac{1}{\sqrt{\rho}} = k\rho^{-\frac{1}{2}},$$

for some constant $k$.

**45. (a)** Since the speed of sound is 340 meters/sec = 0.34 km/sec, if $t$ is in seconds,

$$d = f(t) = 0.34t.$$

Thus, when $t = 5$ sec, $d = 0.34 \cdot 5 = 1.7$ km. Other values in the second row of Table 11.1 are calculated in a similar manner.

**Table 11.1**

| Time, $t$ | 5 sec | 10 sec | 1 min | 5 min |
|---|---|---|---|---|
| Distance, $d$ (km) | 1.7 | 3.4 | 20.4 | 102 |
| Area, $A$ (km$^2$) | 9.1 | 36.3 | 1307 | 32685 |

**(b)** Using $d = 0.34t$, we want to calculate $t$ when $d = 200$, so

$$200 = 0.34t$$
$$t = \frac{200}{0.34} = 588.24 \text{ sec} = 9.8 \text{ mins.}$$

**(c)** The values for $A$ are listed in Table 11.1. These were calculated using the fact that the area of a circle of radius $r$ is $A = \pi r^2$. At time $t$, the radius of the circle of people who have heard the explosion is $d = 0.34t$. Thus

$$A = \pi d^2 = \pi(0.34t)^2 = \pi(0.34)^2 t^2 = 0.363t^2.$$

This formula was used to calculate the values of $A$ in Table 11.1.

**(d)** Since the population density is 31 people/km$^2$, the population, $P$, who have heard the explosion is given by

$$P = 31A.$$

Since $A = 0.363t^2$, we have

$$P = 31 \cdot 0.363t^2 = 11.25t^2.$$

So $P = f(t) = 11.25t^2$.

**(e)** The graph of $P = f(t) = 11.25t^2$ is in Figure 11.2.

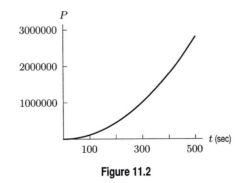

**Figure 11.2**

To find when 1 million people have heard the explosion, we find $t$ when $P = 1,000,000$:

$$1000000 = 11.25t^2$$
$$t^2 = \frac{1000000}{11.25}$$
$$t = \sqrt{\frac{1000000}{11.25}} = 298 \text{ sec} \approx 5 \text{ min.}$$

**49. (a)** The domain of $y = x^{-p}$ is all real numbers except $x = 0$. To find the range we must consider separately the cases when $p$ is even and when $p$ is odd. In the case that $p$ is even, $y$ is always positive. This is because if $p = 2k$ (where k is a positive integer) then

$$y = x^{-p} = \frac{1}{(x^2)^k}$$

and $x^2$ is never negative. Also there is no value of $x$ for which $y$ is equal to zero, because $y = \frac{1}{(x^2)^k}$ can never be zero. Since the function $(x^2)^k$ ranges over all positive numbers, so will the function $y = \frac{1}{(x^2)^k}$. Thus if $p$ is even the range is all positive numbers.

If $p$ is odd then we again note that zero is not in the range. We can rewrite

$$y = x^{-p} = \frac{1}{x^p}.$$

The range of the function $x^p$ is all real numbers. Therefore, the range of $y = \frac{1}{x^p}$ will be the range of the function $x^p$ excluding 0, or all real numbers except 0.

**(b)** If $p$ is even we again write $p = 2k$ (where $k$ is a positive integer) and

$$y = x^{-p} = \frac{1}{(x^2)^k}.$$

Now we note that $y$ is symmetric with respect to the $y$-axis, since

$$\frac{1}{((-x)^2)^k} = \frac{1}{((-1)^2(x^2))^k}$$
$$= \frac{1}{(x^2)^k}$$

If $p$ is odd,

$$\frac{1}{(-x)^p} = \frac{1}{(-1)^p x^p} = -\frac{1}{x^p},$$

so $y(-x) = -y(x)$. Thus, when $p$ is odd, $y = x^{-p}$ is symmetric with respect to the origin.

**(c)** When $p$ is even,

$$x^{-p} \to \infty \qquad \text{as} \qquad x \to 0^+$$
$$x^{-p} \to \infty \qquad \text{as} \qquad x \to 0^-.$$

Thus, values of the function show the same pattern on each side of the $y$-axis – a property we would certainly expect for a function with even symmetry.

When $p$ is odd, we found in part (b) that $y = x^{-p}$ is symmetric about the origin. Therefore, if $y \to +\infty$ as $x \to 0^+$ we would expect $y \to -\infty$ as $x \to 0^-$. This behavior is consistent with the behavior of $y = 1/x$ and $y = 1/x^3$.

**(d)** Again we will write

$$y = x^{-p} = \frac{1}{x^p}.$$

When $x$ is a large positive number, $x^p$ is a large positive number, so its reciprocal is a small positive number. If $p$ is even (and thus the function is symmetric with respect to the $y$-axis) then $y$ will be a small positive number for large negative values of $x$. If $p$ is an odd number (and thus $y$ is symmetric with respect to the origin) then $y$ will be a small negative number for large negative values of $x$.

## Solutions for Section 11.2

### Exercises

**1.** Since $5^x$ is not a power function, this is not a polynomial.

**5.** This is not a polynomial because $2e^x$ is not a power function.

**9.** $y = (x + 4)(2x - 3)(5 - x) = -2x^3 + 5x^2 + 37x - 60$ is a third-degree polynomial with four terms. Its long-run behavior is that of $y = -2x^3$: as $x \to -\infty, y \to +\infty$, as $x \to +\infty, y \to -\infty$.

## Problems

**13.** The window $-10 \leq x \leq 10$ by $-20 \leq y \leq 20$ gives a reasonable picture of both functions. See Figure 11.3. The functions cross the $x$-axis in the same places, which indicates the zeros of $f$ and $g$ are the same. The $y$-intercepts are different, since $f(0) = -5$ and $g(0) = 10$. In addition, the end behaviors of the functions differ. The function $g$ has been flipped about the $x$-axis by the negative coefficient of $x^3$.

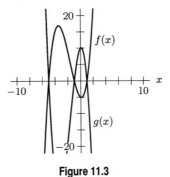

**Figure 11.3**

**17.** Here are the viewing windows used to create these figures. Your viewing windows may differ to a certain extent and still give similar-looking graphs.

**(a)** $-3 \leq x \leq -1, -5 \leq y \leq 5$
**(b)** $-3 \leq x \leq 4, -35 \leq y \leq 15$
**(c)** $1.25 \leq x \leq 2.35, -0 \leq y \leq 6$
**(d)** $-8 \leq x \leq 8, -50 \leq y \leq 2000$

**21.** We have $-20 \leq x \leq 20$, $f(-20) \leq f(20)$ or $-7600 \leq y \leq 8400$.

**25.** **(a)** The graph of the function in the suggested window is shown in Figure 11.4.
**(b)** At $t = 0$ (when Liddleville was founded), the population was 1 hundred people.
**(c)** The $t$-intercept for $t > 0$ will show when the population was zero. This occurs at $t \approx 7.54$. Thus, Liddleville's population reached zero in July of 1897.
**(d)** The graph of $y$ has a peak at $t \approx 3.12$. The population at that point is $\approx 10.1$ hundred. So the maximum population was $\approx 1010$ in February of 1893.
**(e)** Population predicted is $-1.157$ hundred. Since the actual population cannot be negative, we see the model does not predict well at $t = 8$.

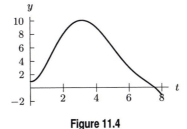

**Figure 11.4**

**29.** Yes. For the sake of illustration, suppose $f(x) = x^2 + x + 1$, a second-degree polynomial. Then

$$f(g(x)) = (g(x))^2 + g(x) + 1$$
$$= g(x) \cdot g(x) + g(x) + 1.$$

Since $f(g(x))$ is formed from products and sums involving the polynomial $g$, the composition $f(g(x))$ is also a polynomial. In general, $f(g(x))$ will be a sum of powers of $g(x)$, and thus $f(g(x))$ will be formed from sums and products involving the polynomial $g(x)$. A similar situation holds for $g(f(x))$, which will be formed from sums and products involving the polynomial $f(x)$. Thus, either expression will yield a polynomial.

**33.** **(a)** Substituting $x = 0.5$ into $p$, we have

$$p(0.5) = 1 - 0.5 + 0.5^2 - 0.5^3 + 0.5^4 - 0.5^5 \approx 0.65625.$$

Since $f(0.5) = 2/3 = 0.6666....$, the approximation is accurate to 2 decimal places.

**(b)** We have $p(1) = 1 - 1 + 1 - 1 + 1 - 1 = 0$, but $f(1) = 0.5$. Thus $p(1)$ is a poor approximation to $f(1)$.

**(c)** See Figure 11.5. The two graphs are difficult to tell apart for $-0.5 \leq x \leq 0.5$, but for $x$ but outside this region the fit is not good.

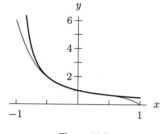

**Figure 11.5**

# Solutions for Section 11.3

## Exercises

**1.** Zeros occur where $y = 0$, which we can find by factoring:

$$x^3 + 7x^2 + 12x = 0$$
$$x(x^2 + 7x + 12) = 0$$
$$x(x + 4)(x + 3) = 0.$$

Zeros are at $x = 0$, $x = -4$, and $x = -3$.

**5.** The graph of $h$ shows zeros at $x = 0$, $x = 3$, and a multiple zero at $x = -2$. Thus

$$h(x) = x(x + 2)^2(x - 3).$$

Check by multiplying and gathering like terms.

**9.** This polynomial must be of fourth (or higher even-powered) degree, so either (but not both) of the zeros at $x = 2$ or $x = 5$ could be doubled. One possible formula is $y = k(x + 2)(x - 2)^2(x - 5)$. Solving for $k$ gives

$$k(0 + 2)(0 - 2)^2(0 - 5) = 5$$
$$-40k = 5$$
$$k = -\frac{1}{8},$$

so

$$y = -\frac{1}{8}(x + 2)(x - 2)^2(x - 5).$$

Another possible formula is $y = k(x + 2)(x - 2)(x - 5)^2$. Solving for $k$ gives

$$k(0 + 2)(0 - 2)(0 - 5)^2 = 5$$
$$-100k = 5$$
$$k = -\frac{1}{20},$$

so

$$y = -\frac{1}{20}(x + 2)(x - 2)(x - 5)^2.$$

There are other possible polynomials, but all are of degree higher than 4, so these are the simplest.

## Problems

**13. (a)** The viewing window $-6 \leq x \leq 2, -10 \leq y \leq 10$ indicates that the zeros of $f$ occur at approximately $x = -5, x = -1, x = 1/2$, and $x = 1$. We therefore guess that

$$f(x) = (x+5)(x+1)(2x-1)(x-1).$$

We check this by expanding the product to obtain $f(x)$.

**(b)** The window $-7 \leq x \leq 2, -150 \leq y \leq 10$ clearly shows all the turning points of $f$.

**17.** Since all the three points fall on a horizontal line, the constant function $f(x) = 1$ (degree zero) is the only polynomial of degree $\leq 2$ to satisfy the given conditions.

**21.** The points $(-3, 0)$ and $(1, 0)$ indicate two zeros for the polynomial. Thus, the polynomial must be of at least degree 2. We could let $p(x) = k(x+3)(x-1)$ as in the previous problems, and then use the point $(0, -3)$ to solve for $k$. An alternative method would be to let $p(x)$ be of the form

$$p(x) = ax^2 + bx + c$$

and solve for $a$, $b$, and $c$ using the given points.

The point $(0, -3)$ gives

$$a \cdot 0 + b \cdot 0 + c = -3,$$
$$\text{so} \quad c = -3.$$

Using $(1, 0)$, we have

$$a(1)^2 + b(1) - 3 = 0$$
$$\text{which gives} \quad a + b = 3.$$

The point $(-3, 0)$ gives

$$a(-3)^2 + b(-3) - 3 = 0$$
$$9a - 3b = 3$$
$$\text{or} \quad 3a - b = 1.$$

From $a + b = 3$, substitute

$$a = 3 - b$$

into

$$3a - b = 1.$$

Then

$$3(3 - b) - b = 1$$
$$9 - 3b - b = 1$$
$$-4b = -8$$
$$\text{so} \quad b = 2.$$

Then $a = 3 - 2 = 1$. Therefore,

$$p(x) = x^2 + 2x - 3$$

is the polynomial of least degree through the given points.

**25.** To obtain the flattened effect of the graph near $x = 0$, let $x = 0$ be a multiple zero (of odd multiplicity). Thus, a possible choice would be $f(x) = kx^3(x+1)(x-2)$ for $k > 0$.

**29.** We see that $h$ has zeros at $x = -2$, $x = -1$ (a double zero), and $x = 1$. Thus, $h(x) = k(x+2)(x+1)^2(x-1)$. Then $h(0) = (2)(1)^2(-1)k = -2k$, and since $h(0) = -2, -2k = -2$ and $k = 1$. Thus,

$$h(x) = (x+2)(x+1)^2(x-1)$$

is a possible formula for $h$.

**33.** $y = 4x^2 - 1 = (2x - 1)(2x + 1)$, which implies that $y = 0$ for $x = \pm\frac{1}{2}$.

**37.** $y = 4x^2 + 1 = 0$ implies that $x^2 = -\frac{1}{4}$, which has no solutions. There are no real zeros.

**41.** **(a)** $V(x) = x(6 - 2x)(8 - 2x)$
   **(b)** Values of $x$ for which $V(x)$ makes sense are $0 < x < 3$, since if $x < 0$ or $x > 3$ the volume is negative.
   **(c)** See Figure 11.6.

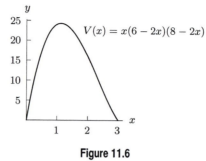

**Figure 11.6**

   **(d)** Using a graphing calculator, we find the peak between $x = 0$ and $x = 3$ to occur at $x \approx 1.13$. The maximum volume is $\approx 24.26 \text{ in}^3$.

**45.** The domain is $x \geq c$ and $a \leq x \leq b$. Taking the hint, we see that the function $y = (x - a)(x - b)(x - c)$ has zeros at $a, b, c$ and long-run behavior like $y = x^3$. Thus, the graph of $y = (x - a)(x - b)(x - c)$ looks something like the graph in the figure (though of course the zeros may be spaced differently). We see that $y < 0$ on the intervals $x < a$ and $b < x < c$. Thus, the function $y = \sqrt{(x - a)(x - b)(x - c)}$ is undefined on these intervals, but is defined everywhere else.

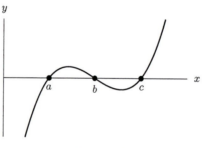

**Figure 11.7**

## Solutions for Section 11.4

### Skill Refresher

**S1.** $\dfrac{6}{y} + \dfrac{7}{y^3} = \dfrac{6y^2 + 7}{y^3}$

**S5.**
$$\frac{5}{(x - 2)^2(x + 1)} - \frac{18}{(x - 2)} = \frac{5 - 18(x - 2)(x + 1)}{(x - 2)^2(x + 1)}$$
$$= \frac{5 - 18x^2 + 18x + 36}{(x - 2)^2(x + 1)0}$$
$$= \frac{-18x^2 + 18x + 41}{(x - 2)^2(x + 1)}$$

**S9.** $\dfrac{x^{-1} + x^{-2}}{1 - x^{-2}} = \dfrac{\dfrac{1}{x} + \dfrac{1}{x^2}}{1 - \dfrac{1}{x^2}} = \dfrac{\dfrac{x + 1}{x^2}}{\dfrac{x^2 - 1}{x^2}} = \dfrac{x + 1}{x^2} \cdot \dfrac{x^2}{x^2 - 1} = \dfrac{x + 1}{(x + 1)(x - 1)} = \dfrac{1}{x - 1}.$

## Exercises

**1.** This is a rational function, and it is already in the form of one polynomial divided by another.

**5.** This is not a rational function, as we cannot put it in the form of one polynomial divided by another, since $\sqrt{x} + 1$ is not a polynomial.

**9.** We have

$$\lim_{x \to \infty} (3x^{-2} + 5x + 7) = \lim_{x \to \infty} (3/x^2 + 5x + 7) = \infty.$$

**13.** As $x \to \pm\infty$, $1/x \to 0$, so $f(x) \to 1$. Therefore $y = 1$ is the horizontal asymptote.

## Problems

**17.** Note: There are many examples to fit these descriptions. Some choices are:

$$\text{Even: } f(x) = \frac{x^2}{x^2 + 1} \qquad \text{Odd: } f(x) = \frac{x^3}{x^2 + 1} \qquad \text{Neither: } f(x) = \frac{x + 1}{x - 1}.$$

If $f(x)$ is even, then $f(-x) = f(x)$. This will be true if and only if both $p(x)$ and $q(x)$ are even or both are odd. If one is even and one is odd, then and only then will $f(x)$ be odd. If one is neither then $f(x)$ is neither.

**21. (a)** Adding $x$ kg of copper increases both the amount of copper and the total amount of alloy. Originally there are 3 kg of copper and 12 kg of alloy. Adding $x$ kg of copper results in a total of $(3 + x)$ kg of copper and $(12 + x)$ kg of alloy. Thus, the new concentration is given by

$$f(x) = \frac{3 + x}{12 + x}.$$

**(b)** (i) $f\left(\frac{1}{2}\right) = \frac{3 + \frac{1}{2}}{12 + \frac{1}{2}} = \frac{\frac{7}{2}}{\frac{25}{2}} = \frac{7}{25} = 28\%$. Thus, adding one-half kilogram copper results in an alloy that is 28% copper.

(ii) $f(0) = \frac{3}{12} = \frac{1}{4} = 25\%$. This means that adding no copper results in the original alloy of 25% copper.

(iii) $f(-1) = \frac{2}{11} \approx 18.2\%$. This could be interpreted as meaning that the removal of 1 kg copper (corresponding to $x = -1$) results in an alloy that is about 18.2% copper.

(iv) Let $y = f(x) = \frac{3 + x}{12 + x}$. Then, multiplying both sides by the denominator we have

$$(12 + x)y = 3 + x$$
$$12y + xy = 3 + x$$
$$xy - x = 3 - 12y$$
$$x(y - 1) = 3 - 12y$$
$$x = \frac{3 - 12y}{y - 1}$$

and so

$$f^{-1}(x) = \frac{3 - 12x}{x - 1}$$

Using this formula, we have $f^{-1}\left(\frac{1}{2}\right) = \frac{-3}{-\frac{1}{2}} = 6$. This means that you must add 6 kg copper in order to obtain an alloy that is $\frac{1}{2}$, or 50%, copper. (You can check this by finding $f(6) = \frac{9}{18} = \frac{1}{2}$).

(v) $f^{-1}(0) = \frac{3}{-1} = -3$. Check: $f(-3) = \frac{0}{9} = 0$. This means that you must remove 3 kg copper to obtain an alloy that is 0% copper, or pure tin.

**(c)** The graph of $y = f(x)$ is in Figure 11.8. The axis intercepts are $(0, 0.25)$ and $(-3, 0)$. The $y$-intercept of $(0, 0.25)$ indicates that with no copper added the concentration is 0.25, or 25%, which is the original concentration of copper in the alloy. The $x$-intercept of $(-3, 0)$ indicates that to make the concentration of copper 0%, we would have to remove 3 kg of copper. This makes sense, as the alloy has only 3 kg of copper to begin with.

**(d)** The graph of $y = f(x)$ on a larger domain is in Figure 11.9.

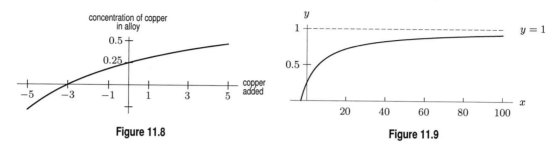

Figure 11.8                    Figure 11.9

The concentration of copper in the alloy rises with the amount of copper added, $x$. However, the graph levels off for large values of $x$, never quite reaching $y = 1 = 100\%$. This is because as more and more copper is added, the concentration gets closer and closer to 100%, but the presence of the 9 kg of tin prevents the alloy from ever becoming 100% pure copper.

**25. (a)** $f(x) = \dfrac{\text{Amount of Alcohol}}{\text{Amount of Liquid}} = \dfrac{x}{x + 5}$

**(b)** $f(7) = \frac{7}{7+5} = \frac{7}{12} \approx 58.333\%$. Also, $f(7)$ is the concentration of alcohol in a solution consisting of 5 gallons of water and 7 gallons of alcohol.

**(c)** $f(x) = 0$ implies that $\dfrac{x}{x + 5} = 0$ and so $x = 0$. The concentration of alcohol is 0% when there is no alcohol in the solution, that is, when $x = 0$.

**(d)** The horizontal asymptote is given by the ratio of the highest-power terms of the numerator and denominator:

$$y = \frac{x}{x} = 1 = 100\%$$

This means that as the amount of alcohol added, $x$, grows large, the concentration of alcohol in the solution approaches 100%.

**29. (a)** $C(x) = 30000 + 3x$

**(b)** $a(x) = \dfrac{C(x)}{x} = \dfrac{30000 + 3x}{x} = 3 + \dfrac{30000}{x}$

**(c)** The graph of $y = a(x)$ is shown in Figure 11.10.

**(d)** The average cost, $a(x)$, approaches \$3 per unit as the number of units grows large. This is because the fixed cost of \$30,000 is averaged over a very large number of goods, so that each good costs only little more than \$3 to produce.

**(e)** The average cost, $a(x)$, grows very large as $x \to 0$, because the fixed cost of \$30000 is being divided among a small number of units.

**(f)** The value of $a^{-1}(y)$ tells us how many units the firm must produce to reach an average cost of \$$y$ per unit. To find a formula for $a^{-1}(y)$, let $y = a(x)$, and solve for $x$. Then

$$y = \frac{30000 + 3x}{x}$$
$$yx = 30000 + 3x$$
$$yx - 3x = 30000$$
$$x(y - 3) = 30000$$
$$x = \frac{30000}{y - 3}.$$

So, we have $a^{-1}(y) = \dfrac{30000}{y - 3}$.

**(g)** We want to evaluate $a^{-1}(5)$, the total number of units required to yield an average cost of \$5 per unit.

$$a^{-1}(5) = \frac{30000}{5-3} = \frac{30000}{2} = 15000.$$

Thus, the firm must produce at least 15,000 units for the average cost per unit to be \$5. The firm must produce at least 15,000 units to make a profit.

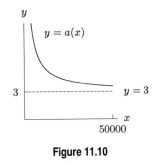

**Figure 11.10**

# Solutions for Section 11.5

## Exercises

**1.** The zero of this function is at $x = 4$. It has vertical asymptotes at $x = \pm 3$. Its long-run behavior is: $y \to 0$ as $x \to \pm\infty$. See Figure 11.11.

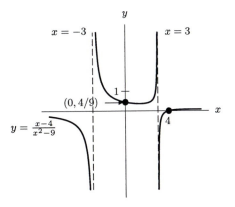

**Figure 11.11**

**5.** Since

$$g(x) = \frac{x^2 - 4}{x^3 + 4x^2} = \frac{(x-2)(x+2)}{x^2(x+4)},$$

the $x$-intercepts are $x = \pm 2$; there is no $y$-intercept; the horizontal asymptote is $y = 0$; the vertical asymptotes are $x = 0, x = -4$.

**9. (a)** See the following table.

| $x$ | $-5$ | $-4.1$ | $-4.01$ | $-4$ | $-3.99$ | $-3.9$ | $-3$ |
|-----|------|--------|---------|------|---------|--------|------|
| $G(x)$ | 10 | 82 | 802 | Undef | $-798$ | $-78$ | $-6$ |

As $x$ approaches $-4$ from the left the function takes on very large positive values. As $x$ approaches $-4$ from the right the function takes on very large negative values.

**(b)**

| $x$ | 5 | 10 | 100 | 1000 |
|-----|---|----|----|------|
| $G(x)$ | 1.111 | 1.429 | 1.923 | 1.992 |

| $x$ | $-5$ | $-10$ | $-100$ | $-1000$ |
|-----|------|-------|--------|---------|
| $G(x)$ | 10 | 3.333 | 2.083 | 2.008 |

For $x > -4$, as $x$ increases, $f(x)$ approaches 2 from below. For $x < -4$, as $x$ decreases, $f(x)$ approaches 2 from above.

**(c)** The horizontal asymptote is $y = 2$. The vertical asymptote is $x = -4$. See Figure 11.12.

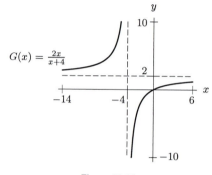

$$G(x) = \frac{2x}{x+4}$$

**Figure 11.12**

## Problems

**13. (a)** To estimate
$$\lim_{x \to 5^+} \frac{x}{5 - x},$$
we consider what happens to the function when $x$ is slightly larger than 5. The numerator is positive and the denominator is negative and is approaching 0 as $x$ approaches 5. We suspect that $\dfrac{x}{5 - x}$ gets more and more negative as $x$ approaches 5 from the right. We can also use either a graph or a table of values as in Table 11.2 to estimate this limit. We see that
$$\lim_{x \to 5^+} \frac{x}{5 - x} = -\infty.$$

**Table 11.2**

| $x$ | 5.1 | 5.01 | 5.001 | 5.0001 |
|-----|-----|------|-------|--------|
| $f(x)$ | $-51$ | $-501$ | $-5001$ | $-50001$ |

**(b)** To estimate
$$\lim_{x \to 5^-} \frac{x}{5 - x},$$
we consider what happens to the function when $x$ is slightly smaller than 5. The numerator is positive and the denominator is positive and is approaching 0 as $x$ approaches 5. We suspect that $\dfrac{x}{5 - x}$ gets larger and larger as $x$ approaches 5 from the left. We can also use either a graph or a table of values to estimate this limit. We see that
$$\lim_{x \to 5^-} \frac{x}{5 - x} = +\infty.$$

**17. (a)** $\lim_{x \to \infty} f(x) = \lim_{x \to -\infty} f(x) = 0.$

**(b)** The vertical asymptote is $x = -2$ and we see

$$\lim_{x \to -2^+} f(x) = \infty \quad \text{and} \quad \lim_{x \to -2^-} f(x) = \infty.$$

**21. (a)** The graph of $y = -f(-x) + 2$ will be the graph of $f$ flipped about both the $x$-axis and the $y$-axis and shifted up 2 units. The graph is shown in Figure 11.13.

**(b)** The graph of $y = \frac{1}{f(x)}$ will have vertical asymptotes $x = -1$ and $x = 3$. As $x \to +\infty$, $\frac{1}{f(x)} \to -\frac{1}{2}$ and as $x \to -\infty$, $\frac{1}{f(x)} \to 0$. At $x = 0$, $\frac{1}{f(x)} = \frac{1}{2}$, and as $x \to -1$ from the left, $\frac{1}{f(x)} \to -\infty$; as $x \to -1$ from the right, $\frac{1}{f(x)} \to +\infty$; as $x \to 3$ from the left, $\frac{1}{f(x)} \to +\infty$; and as $x \to 3$ from the right, $\frac{1}{f(x)} \to -\infty$.

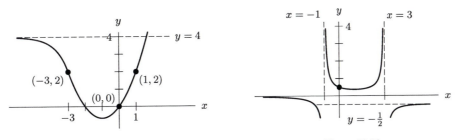

Figure 11.13                         Figure 11.14

**25.** The function $f$ is the transformation of $y = \frac{1}{x}$, so $p = 1$. The graph of $y = \frac{1}{x}$ has been shifted three units to the right and four units up. To find the $y$-intercept, we need to evaluate $f(0)$:

$$f(0) = \frac{1}{-3} + 4 = \frac{11}{3}.$$

To find the $x$-intercepts, we need to solve $f(x) = 0$ for $x$.

$$\text{Thus,} \qquad 0 = \frac{1}{x - 3} + 4,$$
$$-4 = \frac{1}{x - 3},$$
$$-4(x - 3) = 1,$$
$$-4x + 12 = 1,$$
$$-4x = -11,$$
$$\text{so} \qquad x = \frac{11}{4} \quad \text{is the only } x\text{-intercept.}$$

The graph of $f$ is shown in Figure 11.15.

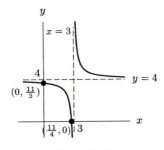

Figure 11.15

**29. (a)** The table indicates translation of $y = 1/x$ because the values of the function are headed in opposite directions near the vertical asymptote.

**(b)** The transformation of $y = 1/x$ given involves a shift to the right by 2 units, so we try

$$y = \frac{1}{x-2}.$$

A check of data from the table shows that we must add $\frac{1}{2}$ to each output of our guess. Thus, a formula would be

$$y = \frac{1}{x-2} + \frac{1}{2}.$$

As a ratio of polynomials, we have

$$y = \frac{1(2)}{(x-2)(2)} + \frac{1(x-2)}{2(x-2)}$$

so

$$y = \frac{x}{2x-4}.$$

**33.**
- Since the graph has a vertical asymptote at $x = 2$, let the denominator be $(x-2)$.
- Since the graph has a zero at $x = -1$, let the numerator be $(x+1)$.
- Since the long–range behavior tends toward $-1$ as $x \to \pm\infty$, the ratio of the leading terms should be $-1$.

So a possible formula is $y = f(x) = -\left(\dfrac{x+1}{x-2}\right)$. You can check that the $y$–intercept is $y = \frac{1}{2}$, as it should be.

**37.** The graph has vertical asymptotes at $x = -1$ and $x = 1$. When $x = 0$, we have $y = 2$ and $y = 0$ at $x = 2$. The graph of $y = \dfrac{(x-2)}{(x+1)(x-1)}$ satisfies each of the requirements, including $y \to 0$ as $x \to \pm\infty$.

**41.** In order to create a hole in the graph of $y = x^3$ at $(2, 8)$, we use the factor $\dfrac{x-2}{x-2}$. Multiplying by this factor is the same as multiplying by 1, except at $x = 2$, where the factor is undefined. So, our function is

$$h(x) = x^3\frac{(x-2)}{(x-2)} = \frac{x^4 - 2x^3}{x-2}.$$

# Solutions for Section 11.6

## Exercises

**1.** The function is exponential, because $p(x) = (5^x)^2 = 5^{2x} = (5^2)^x = 25^x$.

**5.** The function fits an exponential, because $r(x) = 2 \cdot 3^{-2x} = 2(3^{-2})^x = 2(\frac{1}{9})^x$.

**9. (a)**

Table 11.3

| $x$ | $f(x)$ | $g(x)$ |
|---|---|---|
| $-3$ | 1/27 | $-27$ |
| $-2$ | 1/9 | $-8$ |
| $-1$ | 1/3 | $-1$ |
| 0 | 1 | 0 |
| 1 | 3 | 1 |
| 2 | 9 | 8 |
| 3 | 27 | 27 |

**(b)** As $x \to -\infty$, $f(x) \to 0$. For $f$, large negative values of $x$ result in small $f(x)$ values because a large negative power of 3 is very close to zero. For $g$, large negative values of $x$ result in large negative values of $g(x)$, because the cube of a large negative number is a larger negative number. Therefore, as $x \to -\infty$, $g(x) \to -\infty$.

As $x \to \infty$, $f(x) \to \infty$ and $g(x) \to \infty$. For $f(x)$, large $x$-values result in large powers of 3; for $g(x)$, large $x$ values yield the cubes of large $x$-values. $f$ and $g$ both climb *fast*, but $f$ climbs faster than $g$ (for $x > 3$).

**13.** As $x \to \infty$, the higher power dominates, so $x^{1.1}$ dominates $x^{1.08}$. The coefficients 1000 and 50 do not change this, so $y = 50x^{1.1}$ dominates.

## Problems

**17. (a)** Let $f(x) = ax + b$. Then $f(1) = a + b = 18$ and $f(3) = 3a + b = 1458$. Solving simultaneous equations gives us $a = 720, b = -702$. Thus $f(x) = 720x - 702$.

**(b)** Let $f(x) = a \cdot b^x$, then

$$\frac{f(3)}{f(1)} = \frac{ab^3}{ab} = b^2 = \frac{1458}{18} = 81.$$

Thus,

$$b^2 = 81$$
$$b = 9 \quad \text{(since } b \text{ must be positive)}$$

Using $f(1) = 18$ gives

$$a(9)^1 = 18$$
$$a = 2.$$

Therefore, if $f$ is an exponential function, a formula for $f$ would be

$$f(x) = 2(9)^x.$$

**(c)** If $f$ is a power function, let $f(x) = kx^p$, then

$$\frac{f(3)}{f(1)} = \frac{k(3)^p}{k(1)^p} = (3)^p$$

and

$$\frac{f(3)}{f(1)} = \frac{1458}{18} = 81.$$

Thus,

$$3^p = 81 \quad \text{so} \quad p = 4.$$

Solving for $k$, gives

$$18 = k(1^4) \quad \text{so} \quad k = 18.$$

Thus, a formula for $f$ is

$$f(x) = 18x^4.$$

**21.** Note: $\frac{5}{7} > \frac{9}{16} > \frac{3}{8} > \frac{3}{11}$, and we know that for $x > 1$, the higher the exponent, the more steeply the graph climbs, so

$$A \text{ is } kx^{5/7}, \quad B \text{ is } kx^{9/16}, \quad C \text{ is } kx^{3/8}, \quad D \text{ is } kx^{3/11}.$$

**25.** Since $x^8$ is much larger than $x^2 + 5$ for large $x$ (either positive or negative), the ratio tends to zero. Thus, $y \to 0$ as $x \to \infty$ or $x \to -\infty$.

**29.** Multiplying out the numerator gives a polynomial with highest term $x^3$, which dominates the $x^2$ in the denominator. Thus, $y \to x$ as $x \to \infty$ or $x \to -\infty$. So $y \to \infty$ as $x \to \infty$, and $y \to -\infty$ as $x \to -\infty$.

**33.** As $x \to \infty$, the value of $e^x \to \infty$ and $e^{-x} \to 0$. As $x \to -\infty$, the value of $e^{-x} \to \infty$ and $e^x \to 0$. Thus, $y \to \infty$ as $x \to \infty$ and $y \to -\infty$ as $x \to -\infty$.

**37.** The trigonometric function should oscillate, or in other words, the function values should move periodically back and forth between two extremes. It seems $f(x)$ best displays this behavior. The graph in Figure 11.16 shows the points for $f$ from Table 11.21 in the problem. One possible curve has been dashed in. We can recognize the curve as having the same shape as the sine function. The amplitude is 2 (the curve only varies 2 units up or down from the central value of 4). It is raised 4 units from the $x$-axis. Also, the period is 4 because the curve makes one full cycle in the space of 4 units, which tells us that the frequency is $\dfrac{2\pi}{4} = \dfrac{\pi}{2}$. A formula for the curve in Figure 11.16 could be

$$f(x) = 2\sin\left(\frac{\pi}{2}x\right) + 4.$$

(Note: Answers are not unique!)

The exponential function should take the form $y = a \cdot b^x$. Since neither $a$ nor $b$ can be zero, the function cannot pass through the point $(0, 0)$. Therefore, $g(x)$ cannot be exponential. Try $h$ as the exponential function. Figure 11.17 shows the points plotted from $h(x)$ with the curve dashed in.

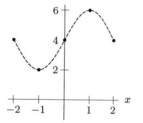

**Figure 11.16**: $f(x)$ best fits a trigonometric function

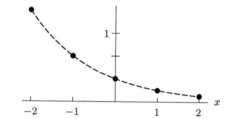

**Figure 11.17**: $h(x)$ could be an exponential function

Rewriting $h(0) = 0.33$ as $h(0) = \frac{1}{3}$ gives $a = \frac{1}{3}$. Thus, using any other point, say $(1, 0.17)$, we have

$$0.17 = \left(\frac{1}{3}\right) b$$

$$b \approx 0.50 = \frac{1}{2}.$$

We find a possible formula to be

$$h(x) = \frac{1}{3}\left(\frac{1}{2}\right)^x.$$

The power function is left for $g$. A power function takes the form $y = k \cdot x^p$. Solving for $k$ and $p$, we find

$$g(x) = -\frac{5}{2}x^3.$$

We see that the data for $g$ satisfy this formula.

# Solutions for Section 11.7

## Exercises

**1.** $f(x) = ax^p$ for some constants $a$ and $c$. Since $f(1) = 1 = a(1)^p$, it follows that $a = 1$. Also, $f(2) = 2^p = c$. Solving for $p$ we have $p = \ln c / \ln 2$. Thus, $f(x) = x^{\ln c / \ln 2}$

**5. (a)** Regression on a calculator returns the power function $f(x) = 201.353x^{2.111}$, where $f(x)$ represents the total dry weight (in grams) of a tree having an $x$ cm diameter at breast height.

**(b)** Using our regression function, we obtain $f(20) = 201.535(20)^{2.111} = 112,313.62$ gm.

**(c)** Solving $f(x) = 201.353x^{2.111} = 100,000$ for $x$ we get

$$x^{2.111} = \frac{100,000}{201.353}$$

$$x = \left(\frac{100,000}{201.353}\right)^{1/2.111} = 18.930 \text{ cm.}$$

**9.** The slope of this line is $m = \frac{y_2 - y_1}{x_2 - x_1} = \frac{3}{2}$. The vertical intercept is 0, thus $y = \frac{3}{2}x$.

## Problems

**13. (a)** The function $y = -83.039 + 61.514x$ gives a superb fit, with correlation coefficient $r = 0.99997$.

**(b)** When the power function is plotted for $2 \le x \le 2.05$, it resembles a line. This is true for most of the functions we have studied. If you zoom in close enough on any given point, the function begins to resemble a line. However, for other values of $x$ (say, $x = 3, 4, 5 \ldots$), the fit no longer holds.

**17. (a)**

| Recliner price ($) | 399 | 499 | 599 | 699 | 799 |
|---|---|---|---|---|---|
| Demand (recliners) | 62 | 55 | 47 | 40 | 34 |
| Revenue ($) | 24,738 | 27,445 | 28,153 | 27,960 | 27,166 |

**(b)** Using quadratic regression, we obtain the following formula for revenue, $R$, as a function of the selling price, $p$: $R(p) = -0.0565p^2 + 72.9981p + 4749.85$.

**(c)** By using a graphing calculator to zoom in, we see that the price which maximizes revenue is about $646. The revenue generated at this price is about $28,349.

**21. (a)** See Figure 11.18.

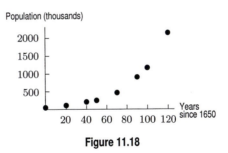

**Figure 11.18**

**(b)** Using a calculator or computer, we get $P(t) = 56.108(1.031)^t$. Answers may vary.

**(c)** The 56.108 represents a population of 56,108 people in 1650. Note that this is more than the actual population of 50,400. The growth factor of 1.031 means the rate of growth is approximately 3.1% per year.

**(d)** We find $P(100) = 1194.308$, which is slightly higher than the given data value of 1,170.8.

**(e)** The estimated population, $P(150) = 5510.118$, is higher than the given census population.

**25. (a)** Using $t = 5, 10, \ldots, 50$, yields $y = 0.310t^2 - 12.177t + 144.517$. Answers may vary.

**(b)** Using $t = 55, 60, 65, 70$, yields $y = 3.01t^2 - 348.43t + 10,955.75$. Answers may vary.

**(c)** $f(t) = \begin{cases} y = 0.310t^2 - 12.177t + 144.517 & 5 \le t \le 50 \\ y = 3.01t^2 - 348.43t + 10,955.75 & 55 \le t \le 70 \end{cases}$

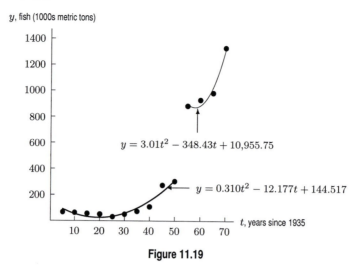

**Figure 11.19**

29. (a) Quadratic is the only choice that increases and then decreases to match the data.
   (b) Using ages of $x = 20, 30, \ldots, 80$, a quadratic function is $y = -34.136x^2 + 3497.733x - 39,949.714$. Answers may vary.
   (c) The value of the function at 37 is $y = -34.136 \cdot 37^2 + 3497.733 \cdot 37 - 39,949.714 = \$42,734$.
   (d) The value of the function for age 10 is $y = -34.136 \cdot 10^2 + 3497.733 \cdot 10 - 39,949.714 = -\$8386$. Answers may vary. Not reasonable, as income is positive. In addition, 10-year olds do not usually work.

33. By plotting $P$ against $D^{3/2}$ we see a straight line with slope$\approx 0.2$. Alternatively, by calculating $P/D^{3/2}$ for each of the planets, we find a common value of approximately 0.2. We see that Kepler's model fits the data reasonably well, with $k \approx 0.2$. Kepler's equation is $P = 0.2d^{3/2}$. Figure 11.20 shows the function $P = 0.2D^{3/2}$ and the data set from Table 11.44, which models the situation well.

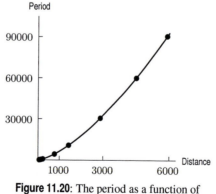

**Figure 11.20**: The period as a function of distance and the function $P = 0.2D^{3/2}$

# Solutions for Chapter 11 Review

## Exercises

**1.** This function represents proportionality to a power.

$$y = \frac{\frac{1}{3}}{2x^7} = \frac{1}{6x^7} = \left(\frac{1}{6}\right) x^{(-7)}.$$

Thus $k = 1/6$ and $p = -7$.

**5.** While $y = 6x^3$ is a power function, when we add two to it, it can no longer be written in the form $y = kx^p$, so this is not a power function.

**9.** Since the graph is symmetric about the $y$-axis, the power function is even.

**13.** Since the graph is symmetric about the origin, the power function has an odd power.

**17.** By multiplying out the expression $(x^2 - 4)(x^2 - 2x - 3)$ and then simplifying the result, we see that

$$y = x^4 - 2x^3 - 7x^2 + 8x + 12$$

which is a fourth-degree polynomial.

**21.** Since $2x^2/x^{-7} = 2x^9$, we can rewrite this polynomial as $y = 2x^9 - 7x^5 + 3x^3 + 2$. Since the leading term of the polynomial is $2x^9$, the value of $y$ goes to infinity as $x \to \infty$. The graph resembles $y = 2x^9$.

**25.** This is not a rational function, as we cannot put it in the form of one polynomial divided by another, since $e^x$ is an exponential function, not a polynomial.

**29. (a)** Since the long-run behavior of $r(x) = p(x)/q(x)$ is given by the ratio of the leading terms of $p$ and $q$, we have

$$\lim_{x \to \infty} \frac{2x + 1}{x - 5} = \lim_{x \to \infty} \frac{2x}{x} = 2.$$

**(b)** We have

$$\lim_{x \to -\infty} \frac{2 + 5x}{6x + 3} = \lim_{x \to -\infty} \frac{5x}{6x} = \frac{5}{6}.$$

## Problems

**33.** Graph (i) looks periodic with amplitude of 2 and period of $2\pi$, so it best corresponds to function J,

$$y = 2\sin(0.5x).$$

Graph (ii) appears to decrease exponentially with a $y$-intercept $< 10$, so it best corresponds to function L,

$$y = 2e^{-0.2x}.$$

Graph (iii) looks like a rational function with two vertical asymptotes, no zeros, a horizontal asymptote at $y = 0$ and a negative $y$-intercept, so it best corresponds to function O,

$$y = \frac{1}{x^2 - 4}.$$

Graph (iv) looks like a logarithmic function with a negative vertical asymptote and $y$-intercept at $(0, 0)$, so it best corresponds to function H,

$$y = \ln(x + 1).$$

**37.** The graph appears to have $x$ intercepts at $x = -\frac{1}{2}, 3, 4$, so let

$$y = k(x + \frac{1}{2})(x - 3)(x - 4).$$

The $y$-intercept is at $(0, 3)$, so substituting $x = 0$, $y = 3$:

$$3 = k(\frac{1}{2})(-3)(-4),$$

which gives    $3 = 6k,$

or    $k = \frac{1}{2}.$

Therefore, $y = \frac{1}{2}(x + \frac{1}{2})(x - 3)(x - 4)$ is a possible formula for $f$.

**41.** We use the position of the "bounce" on the $x$-axis to indicate a multiple zero at that point. Since there is not a sign change at those points, the zero occurs an even number of times.

We have

$$y = k(x + 3)x^2,$$

and using the point $(-1, 2)$ gives

$$2 = k(-1 + 3)(-1)^2 = 2k,$$

so

$$k = 1.$$

Thus, $y = x^2(x + 3)$ is a possible formula.

**45. (a)** The graph indicates the graph of $y = 1/x^2$ has been shifted to the right by 2 and down 1. Thus,

$$y = \frac{1}{(x - 2)^2} - 1$$

is a possible formula for it.

**(b)** The equation $y = 1/(x - 2)^2 - 1$ can be written as

$$y = \frac{-x^2 + 4x - 3}{x^2 - 4x + 4}$$

by obtaining a common denominator and combining terms.

**(c)** The graph has $x$-intercepts when $y = 0$ so the numerator of $y = (-x^2 + 4x - 3)/(x^2 - 4x + 4)$ must equal zero. Then

$$-x^2 + 4x - 3 = 0$$
$$-(x^2 - 4x + 3) = 0$$
$$-(x - 3)(x - 1) = 0,$$

so either $x = 3$ or $x = 1$. The $x$-intercepts are $(1, 0)$ and $(3, 0)$. Setting $x = 0$, we find $y = -\frac{3}{4}$, so $(0, -\frac{3}{4})$ is the $y$-intercept.

**49. (a)** In factored form, $f(x) = x^2 + 5x + 6 = (x + 2)(x + 3)$. Thus, $f$ has zeros at $x = -2$ and $x = -3$. For $g(x) = x^2 + 1$ there are no real zeros.

**(b)** If $r(x) = \frac{f(x)}{g(x)}$, the zeros of $r$ are where the numerator is zero (assuming $g$ is not also zero at those points). Thus, $r$ has zeros $x = -2$ and $x = -3$. There is no vertical asymptote since $g(x)$ is positive for all $x$. As $x \to \pm\infty$, $r$ will behave like $y = \frac{x^2}{x^2} = 1$. Thus, $r(x) \to 1$ as $x \to \pm\infty$. The graph of $y = r(x)$ is shown in Figure 11.21.

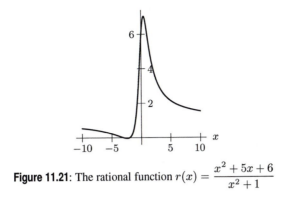

**Figure 11.21**: The rational function $r(x) = \dfrac{x^2 + 5x + 6}{x^2 + 1}$

(c) In fact $s$ does not have a zero near the origin—it does not have a zero anywhere. If $s(x) = 0$ then $g(x) = 0$, which is never true. The function does have two vertical asymptotes, at the zeros of $f$, which are $x = -2$ and $x = 3$. As $x \to \pm\infty$, $s(x) \to 1$.

**53.**
$$f(x) = \frac{p(x)}{q(x)} = \frac{(x+3)(x-2)}{(x+5)(x-7)}$$

**57.** We have $f(x) = k(x + 3)(x - 2)(x - 5)$. Solving for $k$,

$$k(0 + 3)(0 - 2)(0 - 5) = -6$$
$$30k = -6$$
$$k = -\frac{1}{5}.$$

Thus, $f(x) = (-1/5)(x + 3)(x - 2)(x - 5)$.

**61.** The map distance is directly proportional to the actual distance (mileage), because as the actual distance, $x$, increases, the map distance, $d$, also increases.

Substituting the values given into the general formula $d = kx$, we have $0.5 = k(5)$, so $k = 0.1$, and the formula is

$$d = 0.1x.$$

When $d = 3.25$, we have $3.25 = 0.1(x)$ so $x = 32.5$. Therefore, towns which are separated by 3.25 inches on the map are 32.5 miles apart.

**65.** (a) The graph of the function on the suggested window is shown in Figure 11.22. At $x = 0$ (when Smallsville was founded), the population was 5 hundred people.

(b) The $x$-intercept for $x > 0$ will show when the population was zero. This occurs at $x \approx 8.44$. Thus, Smallsville became a ghost town in June of 1908.

(c) There are two peaks on the graph on $0 \le x \le 10$, but the first occurs before $x = 5$ (i.e.,before 1905). The second peak occurs at $x \approx 7.18$. The population at that point is $\approx 7.9$ hundred. So the maximum population was $\approx 790$ in February of 1907.

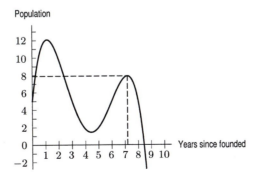

**Figure 11.22**

**69. (a)**

$$p(1) = 1 + 1 + \frac{1^2}{2} + \frac{1^3}{6} + \frac{1^4}{24} + \frac{1^5}{120} \approx 2.71666 \dots .$$

This is accurate to 2 decimal places, since $e \approx 2.718$.

**(b)** $p(5) \approx 91.417$. This is not at all close to $e^5 \approx 148.4$.

**(c)** See Figure 11.23. The two graphs are difficult to tell apart for $-2 \le x \le 2$, but for $x$ much less than $-2$ or much greater than $2$, the fit gets worse and worse.

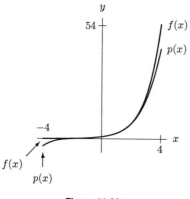

**Figure 11.23**

# CHECK YOUR UNDERSTANDING

**1.** False. The quadratic function $y = 3x^2 + 5$ is not of the form $y = kx^n$, so it is not a power function.

**5.** True. All positive even power functions have an upward opening U shape.

**9.** True. The $x$-axis is an asymptote for $f(x) = x^{-1}$, so the values approach zero.

**13.** False. As $x$ grows very large the exponential decay function $g$ approaches the $x$-axis faster than any power function with a negative power.

**17.** False. For example, the polynomial $x^2 + x^3$ has degree 3 because the degree is the highest power, not the first power, in the formula for the polynomial.

**21.** True. The graph crosses the $y$-axis at the point $(0, p(0))$.

**25.** True. We can write $p(x) = (x - a) \cdot C(x)$. Evaluating at $x = a$ we get $p(a) = (a - a) \cdot C(a) = 0 \cdot C(a) = 0$.

**29.** True. This is the definition of a rational function.

**33.** True. The ratio of the highest degree terms in the numerator and denominator is $2x/x^2 = 2/x$, so for large positive $x$-values, $y$ approaches 0.

**37.** False. The ratio of the highest degree terms in the numerator and denominator is $3x^4/x^2 = 3x^2$. So for large positive $x$-values, $y$ behaves like $y = 3x^2$.

**41.** True. At $x = -4$, we have $f(-4) = (-4 + 4)/(-4 - 3) = 0/(-7) = 0$, so $x = -4$ is a zero.

**45.** False. If $p(x)$ has no zeros, then $r(x)$ has no zeros. For example, if $p(x)$ is a nonzero constant or $p(x) = x^2 + 1$, then $r(x)$ has no zeros.

# Solutions to Skills for Chapter 11

**1.** $\dfrac{3}{5} + \dfrac{4}{7} = \dfrac{3\cdot7 + 4\cdot5}{35} = \dfrac{21+20}{35} = \dfrac{41}{35}$

**5.** $\dfrac{-2}{yz} + \dfrac{4}{z} = \dfrac{-2z + 4yz}{yz^2} = \dfrac{-2 + 4y}{yz} = \dfrac{-2(1-2y)}{yz}$

**9.** $\dfrac{\frac{5}{6}}{15} = \dfrac{5}{6}\cdot\dfrac{1}{15} = \dfrac{1}{18}$

**13.** $\dfrac{4z}{x^2 y} - \dfrac{3w}{xy^4} = \dfrac{4zxy^4 - 3wx^2 y}{x^3 y^5} = \dfrac{xy(4zy^3 - 3wx)}{x^3 y^5} = \dfrac{4y3z - 3wx}{x^2 y^4}$

**17.**
$$\frac{8}{3x^2 - x - 4} - \frac{9}{x+1} = \frac{8}{(x+1)(3x-4)} - \frac{9}{x+1}$$
$$= \frac{8 - 9(3x-4)}{(x+1)(3x-4)}$$
$$= \frac{-27x + 44}{(x+1)(3x-4)}$$

**21.** The second denominator $4r^2 + 6r = 2r(2r+3)$, while the first denominator is $2r+3$. Therefore the common denominator is $2r(2r+3)$. We have:
$$\frac{1}{2r+3} + \frac{3}{4r^2 + 6r} = \frac{1}{2r+3} + \frac{3}{2r(2r+3)}$$
$$= \frac{1\cdot 2r}{2r(2r+3)} + \frac{3}{2r(2r+3)}$$
$$= \frac{2r+3}{2r(2r+3)} = \frac{1}{2r}.$$

**25.** If we factor the number and denominator of the second fraction, we can cancel some terms with the first,
$$\frac{a+b}{2}\cdot\frac{8x+2}{b^2 - a^2} = \frac{a+b}{2}\cdot\frac{2(4x+1)}{(b+a)(b-a)} = \frac{4x+1}{b-a}.$$

**29.** $\dfrac{2a+3}{(a+3)(a-3)}$

**33.** We expand within the first brackets first. Therefore,
$$\frac{[4-(x+h)^2] - [4-x^2]}{h} = \frac{[4-(x^2+2xh+h^2)] - [4-x^2]}{h}$$
$$= \frac{[4-x^2-2xh-h^2] - 4 + x^2}{h} = \frac{-2xh - h^2}{h}$$
$$= -2x - h.$$

**37.** $\dfrac{\dfrac{3}{xy} - \dfrac{5}{x^2 y}}{\dfrac{6x^2 - 7x - 5}{x^4 y^2}} = \dfrac{\dfrac{3x-5}{x^2 y}}{\dfrac{(3x-5)(2x+1)}{y^4 y^2}} = \dfrac{3x-5}{x^2 y}\cdot\dfrac{x^4 y^2}{(3x-5)(2x+1)} = \dfrac{x^2 y}{2x+1}$

**41.** Dividing $2x^3$ into each term in the numerator yields:
$$\frac{26x+1}{2x^3} = \frac{26x}{2x^3} + \frac{1}{2x^3} = \frac{13}{x^2} + \frac{1}{2x^3}.$$

**45.**

$$\frac{\frac{1}{3}x - \frac{1}{2}}{2x} = \frac{\frac{x}{3}}{2x} - \frac{\frac{1}{2}}{2x} = \frac{x}{3} \cdot \frac{1}{2x} - \frac{1}{2} \cdot \frac{1}{2x} = \frac{1}{6} - \frac{1}{4x}$$

**49.** Dividing the denominator $R$ into each term in the numerator yields,

$$\frac{R+1}{R} = \frac{R}{R} + \frac{1}{R} = 1 + \frac{1}{R}.$$

**53.** False

**57.** True

# CHAPTER TWELVE

## Solutions for Section 12.1

### Exercises

1. We can describe an elevation with one number, so this is a scalar.

5. Scalar.

9. See Figure 12.1.

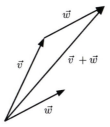

**Figure 12.1**

13. See Figure 12.2.

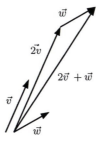

**Figure 12.2**

### Problems

17. **(a)** From Figure 12.3,

$$\text{Distance from Oracle Road} = 5\sin 20^\circ = 1.710 \text{ miles.}$$

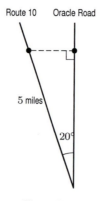

**Figure 12.3**

**(b)** If the distance along Route 10 is $x$ miles, we have

$$x \sin 20° = 2 \text{ miles}$$
$$x = \frac{2}{\sin 20°} = 5.848 \text{ miles}.$$

**21.**

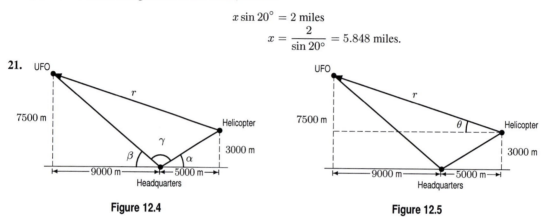

**Figure 12.4**                    **Figure 12.5**

Figure 12.4 shows the headquarters at the origin, and a positive $y$-value as up, and a positive $x$-value as east. To solve for $r$, we must first find $\gamma$:

$$\gamma = 180° - \alpha - \beta$$
$$= 180° - \arctan\frac{3000}{5000} - \arctan\frac{7500}{9000}$$
$$= 109.231°.$$

We now can find $r$ using the Law of Cosines in the triangle formed by the position of the headquarters, the helicopter and the UFO.

In kilometers:

$$r^2 = 34 + 137.250 - 2 \cdot \sqrt{34} \cdot \sqrt{137.250} \cdot \cos\gamma$$
$$r^2 = 216.250$$
$$r = 14.705 \text{ km}$$
$$= 14{,}705 \text{ m}.$$

From Figure 12.5 we see:

$$\tan\theta = \frac{4500}{14{,}000}$$
$$\theta = 17.819°.$$

Therefore, the helicopter must fly 14,705 meters with an angle of 17.819° from the horizontal.

**25.** The vector $\vec{v} + \vec{w}$ is equivalent to putting the vectors $\overrightarrow{OA}$ and $\overrightarrow{AB}$ end-to-end as shown in Figure 12.6; the vector $\vec{w} + \vec{v}$ is equivalent to putting the vectors $\overrightarrow{OC}$ and $\overrightarrow{CB}$ end-to-end. Since they form a parallelogram, $\vec{v} + \vec{w}$ and $\vec{w} + \vec{v}$ are both equal to the vector $\overrightarrow{OB}$, we have $\vec{v} + \vec{w} = \vec{w} + \vec{v}$.

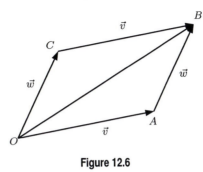

**Figure 12.6**

**29.** According to the definition of scalar multiplication, $1 \cdot \vec{v}$ has the same direction and magnitude as $\vec{v}$, so it is the same as $\vec{v}$.

# Solutions for Section 12.2

## Exercises

**1.** The vector we want is the displacement from $Q$ to $P$, which is given by

$$\overrightarrow{QP} = (1 - 4)\vec{i} + (2 - 6)\vec{j} = -3\vec{i} - 4\vec{j}.$$

**5.** $4\vec{i} + 2\vec{j} - 3\vec{i} + \vec{j} = \vec{i} + 3\vec{j}$.

**9.** $\|\vec{v}\| = \sqrt{1^2 + (-1)^2 + 3^2} = \sqrt{11} \approx 3.317$

## Problems

**13.** The velocity of the ship in still water is $10\vec{j}$ knots and the velocity of the current is $-5\vec{i}$ since the current is east to west. The velocity of the ship is $-5\vec{i} + 10\vec{j}$ knots.

**17.** Figure 12.7 shows the vector $\vec{w}$ redrawn to show that it is perpendicular to the displacement vector $\overrightarrow{PQ}$, which lies along the dotted line. Thus, the angle is $90°$ or $\pi/2$.

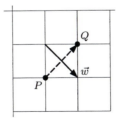

**Figure 12.7**

**21.** We need to calculate the length of each vector.

$$\|21\vec{i} + 35\vec{j}\| = \sqrt{21^2 + 35^2} = \sqrt{1666} \approx 40.8,$$
$$\|40\vec{i}\| = \sqrt{40^2} = 40.$$

So the first car is faster.

**25. (a)** The velocity, $\vec{v}$, is represented by a vector of length 5 in a northeasterly direction. The vector $\vec{i} + \vec{j}$ points northeast, but has length $\sqrt{1^2 + 1^2} = \sqrt{2}$. Thus,

$$\vec{v} = \frac{5}{\sqrt{2}}(\vec{i} + \vec{j}) = 3.536(\vec{i} + \vec{j})$$

**(b)** The current flows northward, so it is represented by $\vec{c} = 1.2\vec{j}$. The swimmer's velocity relative to the riverbed is

$$\vec{s} = \vec{c} + \vec{v} = 1.2\vec{j} + 3.536(\vec{i} + \vec{j}) = 3.536\vec{i} + 4.736\vec{j}.$$

**29.** We get displacement by subtracting the coordinates of the bottom of the tree, $(2, 4, 0)$, from the coordinates of the squirrel, $(2, 4, 1)$, giving:

$$\text{Displacement} = (2 - 2)\vec{i} + (4 - 4)\vec{j} + (1 - 0)\vec{k} = \vec{k}.$$

## Solutions for Section 12.3

### Exercises

**1.** $\vec{B} = 2\vec{M} = 2(1, 1, 2, 3, 5, 8) = (2, 2, 4, 6, 10, 16).$

**5.** $\vec{K} = \dfrac{\vec{N}}{3} + \dfrac{2\vec{N}}{3} = \dfrac{3\vec{N}}{3} = \vec{N} = (5, 6, 7, 8, 9, 10).$

**9.** Since the components of $\vec{Q}$ represent millions of people, an increase of 120,000 people will increase each component by 0.12. Therefore,

$$\begin{aligned}
\vec{S} &= \vec{Q} + (0.12, 0.12, 0.12, 0.12, 0.12, 0.12) \\
&= (3.51, 1.32, 6.40, 1.31, 1.08, 0.62) + (0.12, 0.12, 0.12, 0.12, 0.12, 0.12) \\
&= (3.63, 1.44, 6.52, 1.43, 1.20, 0.74).
\end{aligned}$$

### Problems

**13.** The total scores are out of 300 and are given by the total score vector $\vec{v} + 2\vec{w}$:

$$\begin{aligned}
\vec{v} + 2\vec{w} &= (73, 80, 91, 65, 84) + 2(82, 79, 88, 70, 92) \\
&= (73, 80, 91, 65, 84) + (164, 158, 176, 140, 184) \\
&= (237, 238, 267, 205, 268).
\end{aligned}$$

To get the scores as a percentage, we divide by 3, giving

$$\frac{1}{3}(237, 238, 267, 205, 268) \approx (79.000, 79.333, 89.000, 68.333, 89.333).$$

**17.** **(a)** See the sketch in Figure 12.8, where $\vec{v}$ represents the first part of the man's walk, and $\vec{w}$ represents the second part. Since the man first walks 5 miles, we know $\|\vec{v}\| = 5$. Since he walks 30° north of east, resolving gives

$$\vec{v} = 5\cos 30°\vec{i} + 5\sin 30°\vec{j} = 4.330\vec{i} + 2.500\vec{j}.$$

For the second leg of his journey, the man walks a distance $x$ miles due east, so $\vec{w} = x\vec{i}$.

**(b)** The vector from finish to start is $-(\vec{v} + \vec{w}) = -(4.330 + x)\vec{i} - 2.500\vec{j}$. This vector is at an angle of 10° south of west. So, using the magnitudes of the sides in the triangle in Figure 12.9:

$$\frac{2.500}{4.330 + x} = \tan(10°) = 0.176$$
$$2.500 = 0.176(4.330 + x)$$
$$x = \frac{2.500 - 0.176 \cdot 4.330}{0.176} = 9.848.$$

This means that $x = 9.848$.

**Figure 12.8**                    **Figure 12.9**

**(c)** The distance from the starting point is $\| -(4.330 + 9.848)\vec{i} - (2.500)\vec{j} \| = \sqrt{14.178^2 + 2.500^2} = 14.397$ miles.

**21.** In an actual video game, our rectangle would be replaced with a more sophisticated graphic (perhaps an airplane or an animated figure). But the principles involved in rotation about the origin are the same, and it will be easier to think about them using rectangles instead of fancy graphics.

We can represent the four corners of the rectangle (before rotation) using the position vectors $\vec{p}_a$, $\vec{p}_b$, $\vec{p}_c$, and $\vec{p}_d$. For instance, the components of $\vec{p}_a$ are $\vec{p}_a = 2\vec{i} + \vec{j}$.

After the rectangle has been rotated, its four corners are given by the position vectors $\vec{q}_a$, $\vec{q}_b$, $\vec{q}_c$, and $\vec{q}_d$. Notice that the lengths of these vectors have not changed; in other words,

$$\|\vec{p}_a\| = \|\vec{q}_a\|, \qquad \|\vec{p}_b\| = \|\vec{q}_b\|, \qquad \|\vec{p}_c\| = \|\vec{q}_c\|, \qquad \text{and} \qquad \|\vec{p}_d\| = \|\vec{q}_d\|.$$

This is because in a rotation the only thing that changes is orientation, not length.

When the rectangle is rotated through a 35° angle, the angle made by corner $a$ increases by 35°. So do the angles made by the other three corners. Letting $\theta$ be the angle made by corner $a$, we have

$$\tan\theta = \frac{1}{2}$$
$$\theta = \arctan 0.5 = 26.565°.$$

This is the direction of the position vector $\vec{p}_a$. After rotation, the angle $\theta$ is given by

$$\theta = 26.565° + 35° = 61.565°.$$

This is the direction of the new position vector $\vec{q}_a$. The length of $\vec{q}_a$ is the same as the length of $\vec{p}_a$ and is given by

$$\|\vec{q}_a\|^2 = \|\vec{p}_a\|^2 = 2^2 + 1^2 = 5,$$

and so $\|\vec{q}_a\| = \sqrt{5}$. Thus, the components of $\vec{q}_a$ are given by

$$\vec{q}_a = (\sqrt{5}\cos 61.565°)\vec{i} + (\sqrt{5}\sin 61.565°)\vec{j}$$
$$= 1.065\vec{i} + 1.966\vec{j}.$$

This process can be repeated for the other three corners. You can see for yourself that the angles made with the origin by the corners $a$, $b$, and $c$, respectively, are $14.036°$, $26.565°$, and $45°$. After rotation, these angles are $49.036°$, $61.565°$, and $80°$. Similarly, the lengths of the position vectors for these three points (both before and after rotation) are $\sqrt{17}$, $\sqrt{20}$, and $\sqrt{8}$. Thus, the final positions of these three points are

$$\vec{q}_b = 2.703\vec{i} + 3.113\vec{j},$$
$$\vec{q}_c = 2.129\vec{i} + 3.933\vec{j},$$
$$\vec{q}_d = 0.491\vec{i} + 2.785\vec{j}.$$

## Solutions for Section 12.4

### Exercises

**1.** $\vec{z} \cdot \vec{a} = (\vec{i} - 3\vec{j} - \vec{k}) \cdot (2\vec{j} + \vec{k}) = 1 \cdot 0 + (-3)2 + (-1)1 = 0 - 6 - 1 = -7.$

**5.** $\vec{a} \cdot \vec{b} = (2\vec{j} + \vec{k}) \cdot (-3\vec{i} + 5\vec{j} + 4\vec{k}) = 0(-3) + 2 \cdot 5 + 1 \cdot 4 = 0 + 10 + 4 = 14.$

**9.** Since $\vec{a} \cdot \vec{b}$ is a scalar and $\vec{a}$ is a vector, the expression is a vector parallel to $\vec{a}$. We have

$$\vec{a} \cdot \vec{b} = (2\vec{j} + \vec{k}) \cdot (-3\vec{i} + 5\vec{j} + 4\vec{k}) = 0(-3) + 2(5) + 1(4) = 14.$$

Thus,

$$(\vec{a} \cdot \vec{b}) \cdot \vec{a} = 14\vec{a} = 14(2\vec{j} + \vec{k}) = 28\vec{j} + 14\vec{k}.$$

### Problems

**13.** We use the dot product to find this angle.

We have $\vec{w} = \vec{i} - \vec{j}$ and $\vec{v} = \vec{i} + 2\vec{j}$ so

$$\vec{w} \cdot \vec{v} = (\vec{i} - \vec{j}) \cdot (\vec{i} + 2\vec{j}) = -1,$$

therefore

$$\vec{w} \cdot \vec{v} = ||\vec{w}|| \cdot ||\vec{v}|| \cos \theta.$$

Since $||\vec{w}|| = \sqrt{5}$ and $||\vec{v}|| = \sqrt{2}$, and $\vec{w} \cdot \vec{v} = -1$, we have

$$-1 = \sqrt{2}\sqrt{5} \cos \theta$$
$$\cos \theta = -\frac{1}{\sqrt{10}}$$
$$\theta = \arccos -\frac{1}{\sqrt{10}} = 108.435°.$$

**17.** Using the dot product, the angle is given by

$$\cos \theta = \frac{(\vec{i} + \vec{j} + \vec{k}) \cdot (\vec{i} - \vec{j} - \vec{k})}{||\vec{i} + \vec{j} + \vec{k}|| \, ||\vec{i} - \vec{j} - \vec{k}||} = \frac{1 \cdot 1 + 1(-1) + 1(-1)}{\sqrt{1^1 + 1^2 + 1^2}\sqrt{1^2 + (-1)^2 + (-1)^2}} = -\frac{1}{3}.$$

So, $\theta = \arccos(-\frac{1}{3}) \approx 1.911$ radians, or $\approx 109.471°$.

**21.** It is clear from the Figure 12.10 that only angle $\angle CAB$ could possibly be a right angle. Subtraction of $x, y$ values for the points gives $\overrightarrow{AB} = 3\vec{i} - \vec{j}$ and $\overrightarrow{AC} = 1\vec{i} + 2\vec{j}$. Taking the dot product yields $\overrightarrow{AB} \cdot \overrightarrow{AC} = (3)(1) + (-1)(2) = 1$. Since this is non-zero, the angle can not be a right angle.

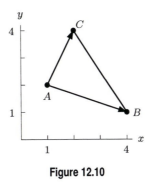

**Figure 12.10**

**25. (a)** We have the price vector $\vec{a} = (3, 2, 4)$. Let the consumption vector $\vec{c} = (c_b, c_e, c_m)$, then $3c_b + 2c_e + 4c_m = 40$ or $\vec{a} \cdot \vec{c} = 40$.

**(b)** Note $\vec{a} \cdot \vec{c}$ is the cost of consuming $\vec{c}$ groceries at Acme Store, so $\vec{b} \cdot \vec{c}$ is the cost of consuming $\vec{c}$ groceries at Beta Mart. Thus $\vec{b} \cdot \vec{c} - \vec{a} \cdot \vec{c} = (\vec{b} - \vec{a}) \cdot \vec{c}$ is the difference in costs between Beta and Acme for the same $\vec{c}$ groceries.

For $\vec{b} - \vec{a}$ to be perpendicular to $\vec{c}$, we must have $(\vec{b} - \vec{a}) \cdot \vec{c} = 0$. Since $\vec{b} - \vec{a} = (0.20, -0.20, 0.50)$, the vector $\vec{b} - \vec{a}$ is perpendicular to $\vec{c}$ if $0.20c_b - 0.20c_e + 0.50c_m = 0$. For example, this occurs when we consume the same number of loaves of bread as dozens of eggs, but no milk.

**(c)** Since $\vec{b} \cdot \vec{c}$ is the cost of groceries at Beta, you might think of $(1/1.1)\vec{b} \cdot \vec{c}$ as the "freshness-adjusted" cost at Beta. Then $(1/1.1)\vec{b} \cdot \vec{c} < \vec{a} \cdot \vec{c}$ means the "freshness-adjusted" cost is cheaper at Beta.

**29. (a)** We have

$$\vec{r} \cdot \vec{w} = (x_1, y_1, x_2, y_2) \cdot (-1, 0, 1, 0)$$
$$= -x_1 + x_2 = x_2 - x_1,$$

so, since $x_2 > x_1$, this quantity represents the width $w$.

**(b)** We have

$$\vec{r} \cdot \vec{h} = (x_1, y_1, x_2, y_2) \cdot (0, -1, 0, 1)$$
$$= -y_1 + y_2 = y_2 - y_1,$$

so, since $y_2 > y_1$, this quantity represents the height $h$.

**(c)** We have

$$2\vec{r} \cdot (\vec{w} + \vec{h}) = 2\vec{r} \cdot \vec{w} + 2\vec{r} \cdot \vec{h}$$
$$= 2w + 2h,$$

where from part (a) $w$ is the width and from part (b) $h$ is the height. Thus, this quantity represents the perimeter $p$.

# Solutions for Section 12.5

## Exercises

**1. (a)** We have

$$5\mathbf{R} = 5 \begin{pmatrix} 3 & 7 \\ 2 & -1 \end{pmatrix} = \begin{pmatrix} 5 \cdot 3 & 5 \cdot 7 \\ 5 \cdot 2 & 5 \cdot -1 \end{pmatrix} = \begin{pmatrix} 15 & 35 \\ 10 & -5 \end{pmatrix}.$$

(b) We have

$$-2\mathbf{S} = -2\begin{pmatrix} 1 & -5 \\ 0 & 8 \end{pmatrix} = \begin{pmatrix} -2\cdot 1 & -2\cdot -5 \\ -2\cdot 0 & -2\cdot 8 \end{pmatrix} = \begin{pmatrix} -2 & 10 \\ 0 & -16 \end{pmatrix}.$$

(c) We have

$$\mathbf{R} + \mathbf{S} = \begin{pmatrix} 3 & 7 \\ 2 & -1 \end{pmatrix} + \begin{pmatrix} 1 & -5 \\ 0 & 8 \end{pmatrix}$$

$$= \begin{pmatrix} 3+1 & 7-5 \\ 2+0 & -1+8 \end{pmatrix} = \begin{pmatrix} 4 & 2 \\ 2 & 7 \end{pmatrix}.$$

(d) Writing $\mathbf{S} - 3\mathbf{R} = \mathbf{S} + (-3)\mathbf{R}$, we first find $-3\mathbf{R}$:

$$-3\mathbf{R} = -3\begin{pmatrix} 3 & 7 \\ 2 & -1 \end{pmatrix} = \begin{pmatrix} -3\cdot 3 & -3\cdot 7 \\ -3\cdot 2 & -3\cdot -1 \end{pmatrix} = \begin{pmatrix} -9 & -21 \\ -6 & 3 \end{pmatrix}.$$

This gives

$$\mathbf{S} + (-3)\mathbf{R} = \begin{pmatrix} 1 & -5 \\ 0 & 8 \end{pmatrix} + \begin{pmatrix} -9 & -21 \\ -6 & 3 \end{pmatrix}$$

$$= \begin{pmatrix} 1-9 & -5-21 \\ 0-6 & 8+3 \end{pmatrix} = \begin{pmatrix} -8 & -26 \\ -6 & 11 \end{pmatrix}.$$

(e) Writing $\mathbf{R} + 2\mathbf{R} + 2(\mathbf{R} - \mathbf{S}) = 5\mathbf{R} + (-2)\mathbf{S}$, we use our answers to parts (a) and (b):

$$5\mathbf{R} + (-2)\mathbf{S} = \begin{pmatrix} 15 & 35 \\ 10 & -5 \end{pmatrix} + \begin{pmatrix} -2 & 10 \\ 0 & -16 \end{pmatrix}$$

$$= \begin{pmatrix} 15-2 & 35+10 \\ 10+0 & -5-16 \end{pmatrix} = \begin{pmatrix} 13 & 45 \\ 10 & -21 \end{pmatrix}.$$

(f) We have

$$k\mathbf{S} = \begin{pmatrix} k\cdot 1 & k\cdot -5 \\ k\cdot 0 & k\cdot 8 \end{pmatrix} = \begin{pmatrix} k & -5k \\ 0 & 8k \end{pmatrix}.$$

5. (a) We have

$$\mathbf{A}\vec{u} = \begin{pmatrix} 2 & 5 & 7 \\ 4 & -6 & 3 \\ 16 & -5 & 0 \end{pmatrix}\begin{pmatrix} 3 \\ 2 \\ 5 \end{pmatrix}$$

$$= \begin{pmatrix} 2\cdot 3 + 5\cdot 2 + 7\cdot 5 \\ 4\cdot 3 - 6\cdot 2 + 3\cdot 5 \\ 16\cdot 3 - 5\cdot 2 + 0\cdot 5 \end{pmatrix} = \begin{pmatrix} 51 \\ 15 \\ 38 \end{pmatrix}.$$

(b) We have

$$\mathbf{B}\vec{v} = \begin{pmatrix} 8 & -6 & 0 \\ 5 & 3 & -2 \\ 3 & 7 & 12 \end{pmatrix} \begin{pmatrix} -1 \\ 0 \\ 3 \end{pmatrix}$$

$$= \begin{pmatrix} 8 \cdot -1 - 6 \cdot 0 + 0 \cdot 3 \\ 5 \cdot -1 + 3 \cdot 0 - 2 \cdot 3 \\ 3 \cdot -1 + 7 \cdot 0 + 12 \cdot 3 \end{pmatrix} = \begin{pmatrix} -8 \\ -11 \\ 33 \end{pmatrix}.$$

(c) Letting $\vec{w} = \vec{u} + \vec{v} = (2, 2, 8)$, we have:

$$\mathbf{A}\vec{w} = \begin{pmatrix} 2 & 5 & 7 \\ 4 & -6 & 3 \\ 16 & -5 & 0 \end{pmatrix} \begin{pmatrix} 2 \\ 2 \\ 8 \end{pmatrix}$$

$$= \begin{pmatrix} 2 \cdot 2 + 5 \cdot 2 + 7 \cdot 8 \\ 4 \cdot 2 - 6 \cdot 2 + 3 \cdot 8 \\ 16 \cdot 2 - 5 \cdot 2 + 0 \cdot 8 \end{pmatrix} = \begin{pmatrix} 70 \\ 20 \\ 22 \end{pmatrix}.$$

Another to work this problem would be to write $\mathbf{A}(\vec{u} + \vec{v})$ as $\mathbf{A}\vec{u} + \mathbf{A}\vec{v}$ and proceed accordingly.

(d) Letting $\mathbf{C} = \mathbf{A} + \mathbf{B}$, we have

$$\mathbf{C} = \begin{pmatrix} 2 & 5 & 7 \\ 4 & -6 & 3 \\ 16 & -5 & 0 \end{pmatrix} + \begin{pmatrix} 8 & -6 & 0 \\ 5 & 3 & -2 \\ 3 & 7 & 12 \end{pmatrix} = \begin{pmatrix} 10 & -1 & 7 \\ 9 & -3 & 1 \\ 19 & 2 & 12 \end{pmatrix}.$$

We can now write $(\mathbf{A} + \mathbf{B})\vec{v}$ as $\mathbf{C}\vec{v}$, and so:

$$\mathbf{C}\vec{v} = \begin{pmatrix} 10 & -1 & 7 \\ 9 & -3 & 1 \\ 19 & 2 & 12 \end{pmatrix} \begin{pmatrix} -1 \\ 0 \\ 3 \end{pmatrix}$$

$$= \begin{pmatrix} 10 \cdot -1 - 1 \cdot 0 + 7 \cdot 3 \\ 9 \cdot -1 - 3 \cdot 0 + 1 \cdot 3 \\ 19 \cdot -1 + 2 \cdot 0 + 12 \cdot 3 \end{pmatrix} = \begin{pmatrix} 11 \\ -6 \\ 17 \end{pmatrix}.$$

(e) From part (a) we have $\mathbf{A}\vec{u} = (51, 15, 38)$, and from part (b) we have $\mathbf{B}\vec{v} = (-8, -11, 33)$. This gives

$$\mathbf{A}\vec{u} \cdot \mathbf{B}\vec{v} = (51, 15, 38) \cdot (-8, -11, 33)$$
$$= 51 \cdot -8 + 15 \cdot -11 + 38 \cdot 33 = 681.$$

(f) We have $\vec{u} \cdot \vec{v} = 3 \cdot -1 + 2 \cdot 0 + 5 \cdot 3 = 12$, and so

$$(\vec{u} \cdot \vec{v})\mathbf{A} = 12\mathbf{A} = 12 \begin{pmatrix} 2 & 5 & 7 \\ 4 & -6 & 3 \\ 16 & -5 & 0 \end{pmatrix} = \begin{pmatrix} 24 & 60 & 84 \\ 48 & -72 & 36 \\ 192 & -60 & 0 \end{pmatrix}.$$

## Problems

**9. (a)** We have

$$s_{\text{new}} = s_{\text{old}} - \underbrace{0.10s_{\text{old}}}_{10\% \text{ infected}}$$

$$= 0.90s_{\text{old}}$$

$$i_{\text{new}} = i_{\text{old}} + \underbrace{0.10s_{\text{old}}}_{10\% \text{ infected}} - \underbrace{0.50i_{\text{old}}}_{50\% \text{ recover}} + \underbrace{0.02r_{\text{old}}}_{2\% \text{ reinfected}}$$

$$= 0.10s_{\text{old}} + 0.50i_{\text{old}} + 0.02r_{\text{old}}$$

$$r_{\text{new}} = r_{\text{old}} + \underbrace{0.50i_{\text{old}}}_{50\% \text{ recover}} - \underbrace{0.02r_{\text{old}}}_{2\% \text{ reinfected}}$$

$$= 0.50i_{\text{old}} + 0.98r_{\text{old}}.$$

Using matrix multiplication, we can rewrite these 3 equations as

$$\begin{pmatrix} s_{\text{new}} \\ i_{\text{new}} \\ r_{\text{new}} \end{pmatrix} = \begin{pmatrix} 0.90 & 0 & 0 \\ 0.10 & 0.50 & 0.02 \\ 0 & 0.50 & 0.98 \end{pmatrix} \begin{pmatrix} s_{\text{old}} \\ i_{\text{old}} \\ r_{\text{old}} \end{pmatrix},$$

and so $\vec{p}_{\text{new}} = \mathbf{T}\vec{p}_{\text{old}}$ where $\mathbf{T} = \begin{pmatrix} 0.90 & 0 & 0 \\ 0.10 & 0.50 & 0.02 \\ 0 & 0.50 & 0.98 \end{pmatrix}$.

**(b)** We have

$$\vec{p_1} = \mathbf{T}\vec{p_0} = \begin{pmatrix} 0.90 & 0 & 0 \\ 0.10 & 0.50 & 0.02 \\ 0 & 0.50 & 0.98 \end{pmatrix} \begin{pmatrix} 2.0 \\ 0.0 \\ 0.0 \end{pmatrix}$$

$$= \begin{pmatrix} 0.9(2) + 0(0) + 0(0) \\ 0.1(2) + 0.5(0) + 0.02(0) \\ 0(2) + 0.5(0) + 0.98(0) \end{pmatrix} = \begin{pmatrix} 1.8 \\ 0.2 \\ 0.0 \end{pmatrix}$$

$$\vec{p_2} = \mathbf{T}\vec{p_1} = \begin{pmatrix} 0.90 & 0 & 0 \\ 0.10 & 0.50 & 0.02 \\ 0 & 0.50 & 0.98 \end{pmatrix} \begin{pmatrix} 1.8 \\ 0.2 \\ 0.0 \end{pmatrix}$$

$$= \begin{pmatrix} 0.9(1.8) + 0(0.2) + 0(0) \\ 0.1(1.8) + 0.5(0.2) + 0.02(0) \\ 0(1.8) + 0.5(0.2) + 0.98(0) \end{pmatrix} = \begin{pmatrix} 1.62 \\ 0.28 \\ 0.10 \end{pmatrix}$$

$$\vec{p_3} = \mathbf{T}\vec{p_2} = \begin{pmatrix} 0.90 & 0 & 0 \\ 0.10 & 0.50 & 0.02 \\ 0 & 0.50 & 0.98 \end{pmatrix} \begin{pmatrix} 1.62 \\ 0.28 \\ 0.10 \end{pmatrix}$$

$$= \begin{pmatrix} 0.9(1.62) + 0(0.28) + 0(0.1) \\ 0.1(1.62) + 0.5(0.28) + 0.02(0.1) \\ 0(1.62) + 0.5(0.28) + 0.98(0.1) \end{pmatrix} = \begin{pmatrix} 1.458 \\ 0.304 \\ 0.238 \end{pmatrix}.$$

**13. (a)** We first find $\vec{v}$ :

$$\vec{v} = \mathbf{A}\vec{u} = \begin{pmatrix} 2 & 1 \\ 3 & 2 \end{pmatrix} \begin{pmatrix} 3 \\ 5 \end{pmatrix} = \begin{pmatrix} 11 \\ 19 \end{pmatrix}.$$

Now, we show that $\vec{u} = \mathbf{A}^{-1}\vec{v}$ :

$$\mathbf{A}^{-1}\vec{v} = \begin{pmatrix} 2 & -1 \\ -3 & 2 \end{pmatrix} \begin{pmatrix} 11 \\ 19 \end{pmatrix} = \begin{pmatrix} 3 \\ 5 \end{pmatrix} = \vec{u}.$$

**(b)** We first find $\vec{v}$ :

$$\vec{v} = \mathbf{A}\vec{u} = \begin{pmatrix} 2 & 1 \\ 3 & 2 \end{pmatrix} \begin{pmatrix} -1 \\ 7 \end{pmatrix} = \begin{pmatrix} 5 \\ 11 \end{pmatrix}.$$

Now, we show that $\vec{u} = \mathbf{A}^{-1}\vec{v}$ :

$$\mathbf{A}^{-1}\vec{v} = \begin{pmatrix} 2 & -1 \\ -3 & 2 \end{pmatrix} \begin{pmatrix} 5 \\ 11 \end{pmatrix} = \begin{pmatrix} -1 \\ 7 \end{pmatrix} = \vec{u}.$$

**(c)** We first find $\vec{v}$ :

$$\vec{v} = \mathbf{A}\vec{u} = \begin{pmatrix} 2 & 1 \\ 3 & 2 \end{pmatrix} \begin{pmatrix} a \\ b \end{pmatrix} = \begin{pmatrix} 2a + b \\ 3a + 2b \end{pmatrix}.$$

Now, we show that $\vec{u} = \mathbf{A}^{-1}\vec{v}$ :

$$\mathbf{A}^{-1}\vec{v} = \begin{pmatrix} 2 & -1 \\ -3 & 2 \end{pmatrix} \begin{pmatrix} 2a + b \\ 3a + 2b \end{pmatrix}$$

$$= \begin{pmatrix} 2(2a + b) - (3a + 2b) \\ -3(2a + b) + 2(3a + 2b) \end{pmatrix}$$

$$= \begin{pmatrix} 4a + 2b - 3a - 2b \\ -6a - 3b + 6a + 4b \end{pmatrix}$$

$$= \begin{pmatrix} a \\ b \end{pmatrix} = \vec{u}.$$

**17. (a)** We have $\mathbf{C} = \begin{pmatrix} 3 & 5 \\ 2 & 4 \end{pmatrix}$ and so

$$\mathbf{C}\vec{u} = \begin{pmatrix} 3 & 5 \\ 2 & 4 \end{pmatrix} \begin{pmatrix} a \\ b \end{pmatrix}$$

$$= \begin{pmatrix} 3a + 5b \\ 2a + 4b \end{pmatrix}$$

$$= \begin{pmatrix} 3a \\ 2a \end{pmatrix} + \begin{pmatrix} 5b \\ 4b \end{pmatrix}$$

$$= a \begin{pmatrix} 3 \\ 2 \end{pmatrix} + b \begin{pmatrix} 5 \\ 4 \end{pmatrix}$$

$$= a\vec{c_1} + b\vec{c_2}.$$

(b) Provided $\mathbf{C}^{-1}$ exists, we can write $\vec{u} = \mathbf{C}^{-1}v$. We can find $\mathbf{C}^{-1}$ using the approach from Problem 14: we have $a = 3, b = 5, c = 2, d = 4$, and so $D = ad - bc = 3(4) - 5(2) = 2$. We have

$$\mathbf{C}^{-1} = \frac{1}{D}\begin{pmatrix} d & -b \\ -c & a \end{pmatrix} = \frac{1}{2}\begin{pmatrix} 4 & -5 \\ -2 & 3 \end{pmatrix} = \begin{pmatrix} 2 & -2.5 \\ -1 & 1.5 \end{pmatrix}.$$

This means that

$$\vec{u} = \mathbf{C}^{-1}\vec{v}$$
$$= \begin{pmatrix} 2 & -2.5 \\ -1 & 1.5 \end{pmatrix}\begin{pmatrix} 2 \\ 5 \end{pmatrix}$$
$$= \begin{pmatrix} 2(2) - 2.5(5) \\ -1(2) + 1.5(5) \end{pmatrix}$$
$$= \begin{pmatrix} -8.5 \\ 5.5 \end{pmatrix}.$$

(c) From parts (a) and (b), we see that

$$\vec{v} = \begin{pmatrix} 2 \\ 5 \end{pmatrix} = \mathbf{C}\vec{u} = \begin{pmatrix} 3 & 2 \\ 5 & 4 \end{pmatrix}\begin{pmatrix} -8.5 \\ 5.5 \end{pmatrix}.$$

Referring to Problem 16, we see that

$$\vec{v} = \begin{pmatrix} 2 \\ 5 \end{pmatrix} = \left(\underbrace{\begin{array}{|c|}3 \\ 2\end{array}}_{\vec{c_1}}\,\underbrace{\begin{array}{|c|}5 \\ 4\end{array}}_{\vec{c_2}}\right)\underbrace{\begin{pmatrix} -8.5 \\ 5.5 \end{pmatrix}}_{\vec{u}}$$
$$= -8.5\underbrace{(3,2)}_{\vec{c_1}} + 5.5\underbrace{(5,4)}_{\vec{c_2}}$$
$$= -8.5\vec{c_1} + 5.5\vec{c_2},$$

which gives $\vec{v}$ as a combination of $\vec{c_1}$ and $\vec{c_2}$.

## Solutions for Chapter 12 Review

### Exercises

**1.** $3\vec{c} = 3(1,1,2) = (3,3,6)$.

**5.** $2\vec{a} - 3(\vec{b} - \vec{c}) = 2(5,1,0) - 3((2,-1,9) - (1,1,2)) = (10,2,0) - 3(1,-2,7) = (7,8,-21)$.

**9.** $-4\vec{i} + 8\vec{j} - 0.5\vec{i} + 0.5\vec{k} = -4.5\vec{i} + 8\vec{j} + 0.5\vec{k}$

**13.** $(5\vec{i} - \vec{j} - 3\vec{k}) \cdot (2\vec{i} + \vec{j} + \vec{k}) = 5 \cdot 2 - 1 \cdot 1 - 3 \cdot 1 = 6$.

**17.** $\vec{a} = \vec{b} = \vec{c} = 3\vec{k}$, $\vec{d} = 2\vec{i} + 3\vec{k}$, $\vec{e} = \vec{j}$, $\vec{f} = -2\vec{i}$.

## Problems

**21.** **(a)** To be parallel, vectors must be scalar multiples. The $\vec{k}$ component of the first vector is 2 times the $\vec{k}$ component of the second vector. So the $\vec{i}$ components of the two vectors must be in a 2:1 ratio, and the same is true for the $\vec{j}$ components. Thus, $4 = 2a$ and $a = 2(a-1)$. These equations have the solution $a = 2$, and for that value, the vectors are parallel.

**(b)** Perpendicular means a zero dot product. So $4a + a(a-1) + 18 = 0$, or $a^2 + 3a + 18 = 0$. Since $b^2 - 4ac = 9 - 4 \cdot 1 \cdot 18 = -63 < 0$, there are no real solutions. This means the vectors are never perpendicular.

**25.** See Figure 12.11, where $\vec{g}$ is the acceleration due to gravity, and $g = \|\vec{g}\|$.

If $\theta = 0$ (the plank is at ground level), the sliding force is $F = 0$.

If $\theta = \pi/2$ (the plank is vertical), the sliding force equals $g$, the force due to gravity.

Therefore, we can guess that $F$ is proportional with $\sin\theta$:

$$F = g\sin\theta.$$

This agrees with the bounds at $\theta = 0$ and $\theta = \pi/2$, and with the fact that the sliding force is smaller than $g$ between 0 and $\pi/2$.

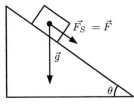

**Figure 12.11**

**29.** The speed of the particle before impact is $v$, so the speed after impact is $0.8v$. If we consider the barrier as being along the $x$-axis (see Figure 12.12), then the $\vec{i}$-component is $0.8v\cos 60° = 0.8v(0.5) = 0.4v$.

Similarly, the $\vec{j}$-component is $0.8v\sin 60° = 0.8v(0.8660) \approx 0.693v$. Thus

$$\vec{v}_{\text{after}} = 0.4v\vec{i} + 0.693v\vec{j}.$$

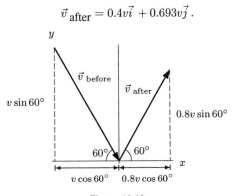

**Figure 12.12**

**33.** If $\vec{x}$ and $\vec{y}$ are two consumption vectors corresponding to points satisfying the same budget constraint, then

$$\vec{p} \cdot \vec{x} = k = \vec{p} \cdot \vec{y}.$$

Therefore we have

$$\vec{p} \cdot (\vec{x} - \vec{y}) = \vec{p} \cdot \vec{x} - \vec{p} \cdot \vec{y} = 0.$$

Thus $\vec{p}$ and $\vec{x} - \vec{y}$ are perpendicular; that is, the difference between two consumption vectors on the same budget constraint is perpendicular to the price vector.

**37.** Using the result of Problem 36, we have $\overrightarrow{AC} = \vec{w} + \vec{n} - \vec{m} = 3\vec{n} - 3\vec{m}$; $\overrightarrow{AB} = \vec{v} + \vec{m} + \vec{n} = 3\vec{m} + \vec{n}$; $\overrightarrow{AD} = \vec{v} + \vec{m} - (\vec{n} - \vec{m}) = 4\vec{m} - \vec{n}$; $\overrightarrow{BD} = (-\vec{n}) - (\vec{n} - \vec{m}) = \vec{m} - 2\vec{n}$.

## CHECK YOUR UNDERSTANDING

1. False. The length $\|0.5\vec{i} + 0.5\vec{j}\| = \sqrt{(0.5)^2 + (0.5)^2} = \sqrt{0.5} \neq 1$.

5. False. The dot product $(2\vec{i} + \vec{j}) \cdot (2\vec{i} - \vec{j}) = 3$ is not zero.

9. True. If $\vec{u} = (u_1, u_2)$ and $\vec{v} = (v_1, v_2)$, then $\vec{u} + \vec{v}$ and $\vec{v} + \vec{u}$ both equal $(u_1 + v_1, u_2 + v_2)$.

13. False. Distance is not a vector. The vector $\vec{i} + \vec{j}$ is the displacement vector from $P$ to $Q$. The distance between the points is the length of the displacement vector: $\|\vec{i} + \vec{j}\| = \sqrt{2}$.

17. True. Both vectors have length $\sqrt{13}$.

21. True. In the subscript, the first number gives the row and the second number gives the column.

25. True. We have

$$A\vec{v} = \begin{pmatrix} 2 & 3 \\ -1 & 5 \end{pmatrix} \begin{pmatrix} 10 \\ 5 \end{pmatrix} = \begin{pmatrix} 2 \cdot 10 + 3 \cdot 5 \\ -10 + 5 \cdot 5 \end{pmatrix} = \begin{pmatrix} 35 \\ 15 \end{pmatrix}.$$

# CHAPTER THIRTEEN

## Solutions for Section 13.1

### Exercises

**1.** Not arithmetic. The differences are 5, 4, 3.

**5.** Arithmetic, with $a = 6$, $d = 3$, so $a_n = 6 + (n-1)3 = 3 + 3n$.

**9.** Not geometric. The ratios of successive terms are $2, 2, \frac{3}{2}$.

**13.** Not geometric, since the ratios of successive terms are $1/4$, $1/4$, $1/2$.

**17.** Geometric, since the ratios of successive terms are all $1/1.2$. Thus, $a = 1$ and $r = 1/1.2$, so $a_n = 1(1/1.2)^{n-1} = 1/(1.2)^{n-1}$.

### Problems

**21.** $a_1 = \cos(\pi) = -1$, $a_2 = \cos(2\pi) = 1$, $a_3 = \cos(3\pi) = -1$, $a_4 = \cos(4\pi) = 1$. This is a geometric sequence.

**25.** Since the first term is 5 and the difference is 10, the arithmetic sequence is given by

$$a_n = 5 + (n-1)10.$$

For $a_n = 5 + (n-1) \cdot 10 > 1000$, we must have

$$10(n-1) > 995$$
$$n - 1 > 99.5$$
$$n > 100.5.$$

The terms of the sequence exceed 1000 when $n \geq 101$.

**29.** Since $a_6 - a_3 = 3d = 9 - 5.7 = 3.3$, we have $d = 1.1$, so $a_1 = a_3 - 2d = 5.7 - 2 \cdot 1.1 = 3.5$. Thus

$$a_5 = 3.5 + (5-1)1.1 = 7.9$$
$$a_{50} = 3.5 + (50-1)1.1 = 57.4$$
$$a_n = 3.5 + (n-1)1.1 = 2.4 + 1.1n.$$

**33.** Since $a_2 = ar = 6$ and $a_4 = ar^3 = 54$, we have

$$\frac{a_4}{a_2} = \frac{ar^3}{ar} = r^2 = \frac{54}{6} = 9,$$

so

$$r = 3.$$

We have $a = 6/r = 6/3 = 2$. Thus

$$a_6 = ar^5 = 2 \cdot (3)^5 = 486$$
$$a_n = ar^{n-1} = 2 \cdot 3^{n-1}.$$

**37. (a)** The growth factor of the population is

$$r = \frac{17.960}{17.613} = 1.0197.$$

Thus, one and two years after 2005, the population is

$$a_1 = 17.960(1.0197) = 18.314$$
$$a_2 = 17.960(1.0197)^2 = 18.675.$$

**(b)** Using the growth factor in part (a), $n$ years after 2005, the population is

$$a_n = 17.960(1.0197)^n.$$

**(c)** The doubling time is the value of $n$ for which

$$a_n = 2 \cdot 17.960$$
$$17.960(1.0197)^n = 2 \cdot 17.960$$
$$(1.0197)^n = 2$$
$$n \ln(1.0197) = \ln 2$$
$$n = \frac{\ln 2}{\ln(1.0197)} = 35.531.$$

Thus the doubling time is about 36.5 years

**41.** Arithmetic, because points lie on a line. The sequence is decreasing, so $d < 0$.

**45.** Since $a_1 = 3$ and $a_n = 2a_{n-1} + 1$, we have $a_2 = 2a_1 + 1 = 7$, $a_3 = 2a_2 + 1 = 15$, $a_4 = 2a_3 + 1 = 31$. Writing out the terms without simplification to try and guess the pattern, we have

$$a_1 = 3$$
$$a_2 = 2 \cdot 3 + 1$$
$$a_3 = 2(2 \cdot 3 + 1) + 1 = 2^2 \cdot 3 + 2 + 1$$
$$a_4 = 2(2^2 \cdot 3 + 2 + 1) + 1 = 2^3 \cdot 3 + 2^2 + 2 + 1.$$

Thus

$$a_n = 2^{n-1} \cdot 3 + 2^{n-2} + 2^{n-3} + \cdots + 2 + 1.$$

**49. (a)** You send the letter to 4 friends; at that stage your name is on the bottom of the list. The 4 friends send it to 4 each, so $4^2$ people have the letter when you are third on the list. These $4^2$ people send it to 4 each, so $4^3$ people have the letter when you are second on the list. By similar reasoning, $4^4$ people have the letter when you are at the top of the list. Thus, you should receive $\$4^4 = \$256$. (The catch is, of course, that someone usually breaks the chain.)
**(b)** By the same reasoning as in part (a), we see that $d_n = 4^n$.

## Solutions for Section 13.2

### Exercises

**1.** This series is not arithmetic, as each term is twice the previous one.

**5.** $\sum\limits_{i=-1}^{5} i^2 = (-1)^2 + 0^2 + 1^2 + 2^2 + 3^2 + 4^2 + 5^2.$

**9.** $\displaystyle\sum_{j=2}^{10}(-1)^j = (-1)^2 + (-1)^3 + (-1)^4 + \cdots + (-1)^{10}.$

**13.** The pattern is $1/2, 2/2, 3/2, \ldots, 8/2$. Thus, a possible solution is

$$\sum_{n=1}^{8}\frac{1}{2}n.$$

**17.** One way to work this problem is to complete the first few values of $a_n$, and then continue the pattern across the row. Then, the values of $S_n$ can be found by summing the values of $a_n$ from left to right. See Table 13.1. We see that $a_1 = 3$ and $d = a_2 - a_1 = 7 - 3 = 4$.

**Table 13.1**

| $n$ | 1 | 2 | 3 | 4 | 5 | 6 | 7 | 8 |
|---|---|---|---|---|---|---|---|---|
| $a_n$ | 3 | 7 | 11 | 15 | 19 | 23 | 27 | 31 |
| $S_n$ | 3 | 10 | 21 | 36 | 55 | 78 | 105 | 136 |

**21.** We use the formula $S_n = 1 + 2 + 3 + \cdots + n = \frac{1}{2}n(n+1)$ with $n = 1000$:

$$S_{1000} = \frac{1}{2} \cdot 1000 \cdot 1001 = 500{,}500.$$

**25.** $\sum_{n=0}^{10}(8 - 4n) = 8 + 4 + 0 + (-4) + \cdots$. This is an arithmetic series with 11 terms and $d = -4$.

$$S_{11} = \frac{1}{2} \cdot 11(2 \cdot 8 + 10(-4)) = -132.$$

**29.** This is an arithmetic series with $a_1 = -3.01$, $n = 35$, and $d = -0.01$. Thus

$$S_{35} = \frac{1}{2} \cdot 35(2(-3.01) + 34(-0.01)) = -111.3.$$

## Problems

**33. (a)** (i) From the table, we have $S_4 = 226.6$, the population of the US in millions 4 decades after 1940, that is, in 1980. Similarly, $S_5 = 248.7$, the population in millions in 1990, and $S_6 = 281.4$, the population in 2000.

(ii) We have $a_2 = S_2 - S_1 = 179.3 - 150.7 = 28.6$; that is, the increase in the US population in millions in the 1950s.

Similarly, $a_5 = S_5 - S_4 = 248.7 - 226.6 = 22.1$, the population increase in millions during the 1980s. In the same way, $a_6 = S_6 - S_5 = 281.4 - 248.7 = 32.7$, the population increase during the 1990s.

(iii) Using the answer to (ii), we have $a_6/10 = 32.7/10 = 3.27$, the average yearly population growth during the 1990s.

(iv) We have

$S_n = $ US population, in millions, $n$ decades after 1940.

$a_n = S_n - S_{n-1} = $ growth in US population in millions, during the $n^{\text{th}}$ decade after 1940.

$a_n/10 = $ Average yearly growth, in millions, during the $n^{\text{th}}$ decade after 1940.

**37.** Expanding both sums, we see

$$\sum_{i=1}^{15}i^3 - \sum_{j=3}^{15}j^3 = (1^3 + 2^3 + 3^3 + 4^3 + \cdots + 15^3) - (3^3 + 4^3 + \cdots + 15^3)$$

$$= 1^3 + 2^3 = 9.$$

**41.** We have $a_1 = 16$ and $d = 32$. At the end of $n$ seconds, the object has fallen

$$S_n = \frac{1}{2}n\left(2 \cdot 16 + (n-1)32\right) = \frac{1}{2}n(32 + 32n - 32) = 16n^2.$$

The height of the object at the end of $n$ seconds is

$$h = 1000 - S_n = 1000 - 16n^2.$$

When the object hits the ground, $h = 0$, so

$$0 = 1000 - 16n^2$$

$$n^2 = \frac{1000}{16} \quad \text{so} \quad n = \pm\sqrt{\frac{1000}{16}} = \pm 7.906 \text{ sec.}$$

Since $n$ is positive, it takes 7.906 seconds, that is, nearly 8 seconds, for the object to hit the ground.

**45. (a)** Since there are 9 terms, we can group them into 4 pairs each totaling 66. The middle or fifth term, $a_5 = 33$, remains unpaired. This means that

$$\text{Sum of series} = 4(66) + 33 = 297.$$

This is the same answer we got by adding directly.

**(b)** We can use our formula derived for the sum of an even number of terms to add the first 8 terms. We have $a_1 = 5$, $n = 8$, and $d = 7$. This gives

$$\text{Sum of first 8 terms} = \frac{1}{2}(8)(2(5) + (8-1)(7)) = 236.$$

Adding the ninth term gives

$$\text{Sum of series} = 236 + 61 = 297.$$

This is the same answer we got by adding directly.

**(c)** Using the method from part (a), we add the first and last terms and obtain $a_1 + a_n$. Notice that $a_n = a_1 + (n-1)d$ and so this expression can be rewritten as $a_1 + a_1 + (n-1)d = 2a_1 + (n-1)d$. We then add the second and next to last terms and obtain $a_2 + a_{n-1}$. We know that $a_2 = a_1 + d$ and that $a_{n-1} = a_1 + (n-2)d$, and so this expression can be rewritten as $a_1 + d + a_1(n-2)d = 2a_1 + (n-1)d$. Continuing in this manner, we see that the sum of each pair is $2a_1 + (n-1)d$. The total number of such terms is given by $\frac{1}{2}(n-1)$. For instance, if there are 9 terms, the total number of pairs is $\frac{1}{2}(9-1) = 4$. Thus, the subtotal of these pairs is given by

$$\text{Subtotal of pairs} = \frac{1}{2}(n-1)(2a_1 + (n-1)d).$$

As you can check for yourself, the unpaired (middle) term is given by

$$\text{Unpaired (middle) term} = a_1 + \frac{1}{2}(n-1)d.$$

For instance, in the case of the series given in the question, the unpaired term is given by

$$5 + \frac{1}{2}(9-1)(7) = 5 + 4(7) = 33.$$

Therefore, the sum of the arithmetic series is given by

$$\text{Sum} = \text{Subtotal of pairs} + \text{Unpaired term}$$
$$= \underbrace{\frac{1}{2}(n-1)(2a_1 + (n-1)d)}_{\text{Subtotal of pairs}} + \underbrace{a_1 + \frac{1}{2}(n-1)d}_{\text{Unpaired term}}.$$

One way to simplify this expression is to first factor out 1/2:

$$\text{Sum} = \frac{1}{2}\left[(n-1)(2a_1 + (n-1)d) + 2a_1 + (n-1)d\right].$$

The bracketed part of this expression involves $(n-1)$ terms equaling $2a_1 + (n-1)d$, plus one more such term, for a total number of $n$ such terms. We have

$$\text{Sum} = \frac{1}{2}n(2a_1 + (n-1)d),$$

which is the same as the formula derived in the text.

Using the method from part (b), we see that since $n$ is odd, $(n-1)$ is even, so we can use the formula derived for an even number of terms to add the first $(n-1)$ terms. Substituting $(n-1)$ for $n$ in our formula, we have

$$\text{Sum of first } (n-1) \text{ terms} = \frac{1}{2}(n-1)(2a_1 + (n-2)d).$$

The total sum is given by

$$\text{Sum} = \text{Sum of first } (n-1) \text{ terms} + n^{\text{th}} \text{ term}.$$

Since the $n^{\text{th}}$ term is given by $a_1 + (n-1)d$, we have

$$\text{Sum} = \underbrace{\frac{1}{2}(n-1)(2a_1 + (n-2)d)}_{\text{Sum of first } (n-1) \text{ terms}} + \underbrace{a_1 + (n-1)d}_{n^{\text{th}} \text{ term}}.$$

To simplify this expression, we first factor out $1/2$, as before:

$$\text{Sum} = \frac{1}{2}\left[(n-1)(2a_1 + (n-2)d) + 2a_1 + 2(n-1)d\right].$$

We can rewrite $2a_1 + 2(n-1)d$ as $2a_1 + 2nd - 2d$, and then as $2a_1 + nd - 2d + nd$, and finally as $2a_1 + (n-2)d + nd$. This gives

$$\text{Sum} = \frac{1}{2}\left[(n-1)(2a_1 + (n-2)d) + 2a_1 + (n-2)d + nd\right].$$

We have $n-1$ terms each equaling $2a_1 + (n-2)d$ plus one more such term plus a term equaling $nd$. This gives a total of $n$ terms equaling $2a_1 + (n-2)d$ plus the $nd$ term:

$$\begin{aligned}
\text{Sum} &= \frac{1}{2}\left[n(2a_1 + (n-2)d) + nd\right] \\
&= \frac{1}{2}n\left[2a_1 + (n-2)d + d\right] \qquad \text{factoring out } n \\
&= \frac{1}{2}n(2a_1 + (n-1)d),
\end{aligned}$$

which is the same answer as before.

# Solutions for Section 13.3

## Exercises

1. Since

$$\sum_{j=5}^{18} = 3 \cdot 2^5 + 3 \cdot 2^6 + 3 \cdot 2^7 + \cdots + 3 \cdot 2^{18},$$

there are $18 - 4 = 14$ terms in the series. The first term is $a = 3 \cdot 2^5$ and the ratio is $r = 2$, so

$$\text{Sum} = \frac{3 \cdot 2^5(1 - 2^{14})}{1 - 2} = 1{,}572{,}768.$$

**5.** This is a geometric series with a first term is $a = 1/125$ and a ratio of $r = 5$. To determine the number of terms in the series, use the formula, $a_n = ar^{n-1}$. Calculating successive terms of the series $(1/125)5^{n-1}$, we get

$$1/125, 1/25, 1/5, 1, 5, 25, 125, 625.$$

Thus we sum the first eight terms of the series:

$$S_8 = \frac{(1/125)(1 - 5^8)}{1 - 5} = \frac{97656}{125} = 781.248.$$

**9.** Yes, $a = 1$, ratio $= -1/2$.

**13.** These numbers are all power of 3, with signs alternating from positive to negative. We need to change the signs on an alternating basis. By raising $-1$ to various powers, we can create the pattern shown. One possible answer is: $\sum_{n=1}^{6}(-1)^{n+1}(3^n)$.

**17.** $\sum_{n=0}^{n=5} 3/(2^n) = 3 + 3/2 + 3/4 + 3/8 + 3/16 + 3/32 = 3(1 - 1/2^6)/(1 - 1/2)) = 189/32$

## Problems

**21. (a)** Let $a_n$ be worldwide oil consumption $n$ years after 2003. Then, $a_1 = 81$, $a_2 = 81(1.012)$, and $a_n = 81(1.012)^{n-1}$. Thus, between 2004 and 2028,

$$\text{Total oil consumption} = \sum_{n=1}^{25} 81(1.012)^{n-1} \text{ billion barrels.}$$

**(b)** Using the formula for the sum of a finite geometric series, we have

$$\text{Total oil consumption} = \frac{81\left(1 - (1.012)^{25}\right)}{1 - 1.012} = 2345.291 \text{ billion barrels.}$$

**25.** The answer to Problem 24 is given by

$$B_{20} = \frac{1000(1 - (1.03)^{20})}{1 - 1.03} = 26,870.37.$$

**(a)** Replacing $1000 by $2000 doubles the answer, giving $53,740.75.
**(b)** Doubling the interest rate to 6% by replacing 1.03 by 1.06 less than doubles the answer, giving $36,785.60.
**(c)** Doubling the number of deposits to 40 by replacing $(1.03)^{20}$ by $(1.03)^{40}$ more than doubles the answer, giving $75,401.26.

## Solutions for Section 13.4

## Exercises

**1.** Yes, $a = 1$, ratio $= -x$.

**5.** Yes, $a = e^x$, ratio $= e^x$.

**9.** Sum $= \dfrac{1}{1 - (-x)} = \dfrac{1}{1 + x}, |x| < 1.$

**13.**

$$\sum_{i=4}^{\infty} \left(\frac{1}{3}\right)^i = \left(\frac{1}{3}\right)^4 + \left(\frac{1}{3}\right)^5 + \cdots = \left(\frac{1}{3}\right)^4 \left(1 + \frac{1}{3} + \left(\frac{1}{3}\right)^2 + \cdots\right) = \frac{\left(\frac{1}{3}\right)^4}{1 - \frac{1}{3}} = \frac{1}{54}.$$

**17.** Since

$$\sum_{i=1}^{\infty} x^{2i} = x^2 + x^4 + x^6 \cdots,$$

the first term is $a = x^2$ and the ratio is $r = x^2$. Since $|x| < 1$,

$$\text{Sum} = \sum_{i=1}^{\infty} x^{2i} = \frac{a}{1-r} = \frac{x^2}{1-x^2}.$$

## Problems

**21.** $0.122222\ldots = 0.1 + \dfrac{2}{100} + \dfrac{2}{1000} + \dfrac{2}{10000} + \dfrac{2}{100000} + \cdots$. Thus,

$$S = 0.1 + \frac{\frac{2}{100}}{1 - \frac{1}{10}} = \frac{1}{10} + \frac{2}{90} = \frac{11}{90}.$$

**25. (a)**

$$P_1 = 0$$
$$P_2 = 250(0.04)$$
$$P_3 = 250(0.04) + 250(0.04)^2$$
$$P_4 = 250(0.04) + 250(0.04)^2 + 250(0.04)^3$$
$$\vdots$$
$$P_n = 250(0.04) + 250(0.04)^2 + \cdots + 250(0.04)^{n-1}$$

**(b)** Factoring our formula for $P_n$, we see that it involves a geometric series of $n - 2$ terms:

$$P_n = 250(0.04) \underbrace{\left[1 + 0.04 + (0.04)^2 + \cdots + (0.04)^{n-2}\right]}_{n-2 \text{ terms}}.$$

The sum of this series is given by

$$1 + 0.04 + (0.04)^2 + \cdots + (0.04)^{n-2} = \frac{1 - (0.04)^{n-1}}{1 - 0.04}.$$

Thus,

$$P_n = 250(0.04)\left(\frac{1 - (0.04)^{n-1}}{1 - 0.04}\right)$$
$$= 10\left(\frac{1 - (0.04)^{n-1}}{1 - 0.04}\right).$$

**(c)** In the long run, that is, as $n \to \infty$, we know that $(0.04)^{n-1} \to 0$, and so

$$P_n = 10\left(\frac{1 - (0.04)^{n-1}}{1 - 0.04}\right) \to 10\left(\frac{1 - 0}{1 - 0.04}\right) = 10.417.$$

Thus, $P_n$ gets closer to 10.417 and $Q_n$ gets closer to 260.42. We'd expect these limits to differ because one is right before taking a tablet and one is right after. We'd expect the difference between them to be exactly 250 mg, the amount of ampicillin in one tablet.

**29.**

$$\text{Present value of first coupon} = \frac{50}{1.04}$$

$$\text{Present value of second coupon} = \frac{50}{(1.04)^2}, \text{etc.}$$

$$\begin{aligned}
\text{Total present value} &= \underbrace{\frac{50}{1.04} + \frac{50}{(1.04)^2} + \cdots + \frac{50}{(1.04)^{10}}}_{\text{coupons}} + \underbrace{\frac{1000}{(1.04)^{10}}}_{\text{principal}} \\
&= \frac{50}{1.04}\left(1 + \frac{1}{1.04} + \cdots + \frac{1}{(1.04)^9}\right) + \frac{1000}{(1.04)^{10}} \\
&= \frac{50}{1.04}\left(\frac{1 - \left(\frac{1}{1.04}\right)^{10}}{1 - \frac{1}{1.04}}\right) + \frac{1000}{(1.04)^{10}} \\
&= 405.545 + 675.564 \\
&= \$1081.11.
\end{aligned}$$

# Solutions for Chapter 13 Review

## Exercises

**1.** We have $n = 18$, $a_1 = 8$, $d = 3$. To find the sum of the series we use the formula $S = (1/2)n(2a_1 + (n-1)d)$. So

$$S_{18} = \frac{18}{2}(2(8) + 17(3)) = 603.$$

To find the $18^{\text{th}}$ term, we use the formula $a_n = a_1 + (n-1)d$. Thus,

$$a_{18} = 8 + 17(3) = 59.$$

**5.** No. Ratio between successive terms is not constant: $\dfrac{x^2/3}{x/2} = \dfrac{2x}{3}$, while $\dfrac{x^3/4}{x^2/3} = \dfrac{3x}{4}$.

**9.** This is an arithmetic series. We have $n = 9$, $a_1 = 7$, $d = 7$. Using our formula, $S_n = (1/2)n(2a_1 + (n-1)d)$, we have

$$S_9 = \frac{9}{2}(2 \cdot 7 + (9-1)7) = 315.$$

## Problems

**13.** We know $S_8 = 108$, $a_1 = 24$ and $n = 8$. We need to find $d$. Substituting in the formula $S = (1/2)n(2a_1 + (n-1)d)$, we get

$$\begin{aligned}
108 &= \frac{1}{2}(8)(2(24) + 7d) \\
108 &= 4(48 + 7d) \\
108 &= 192 + 28d \\
d &= -3.
\end{aligned}$$

He can make 8 rows of cans, starting with 24 cans on the bottom. Each row of cans will have 3 fewer cans than the row underneath it.

**17.**

$$\text{Total present value, in dollars} = 1000 + 1000e^{-0.04} + 1000e^{-0.04(2)} + 1000e^{-0.04(3)} + \cdots$$
$$= 1000 + 1000(e^{-0.04}) + 1000(e^{-0.04})^2 + 1000(e^{-0.04})^3 + \cdots$$

This is an infinite geometric series with $a = 1000$ and $x = e^{(-0.04)}$, and sum

$$\text{Total present value, in dollars} = \frac{1000}{1 - e^{-0.04}} = 25{,}503.33.$$

**21. (a)** Let $h_n$ be the height of the $n^{\text{th}}$ bounce after the ball hits the floor for the $n^{\text{th}}$ time. Then from Figure 13.1,

$$h_0 = \text{height before first bounce} = 10 \text{ feet},$$
$$h_1 = \text{height after first bounce} = 10\left(\frac{3}{4}\right) \text{ feet},$$
$$h_2 = \text{height after second bounce} = 10\left(\frac{3}{4}\right)^2 \text{ feet}.$$

Generalizing this gives

$$h_n = 10\left(\frac{3}{4}\right)^n.$$

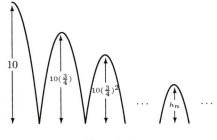

**Figure 13.1**

**(b)** When the ball hits the floor for the first time, the total distance it has traveled is just $D_1 = 10$ feet. (Notice that this is the same as $h_0 = 10$.) Then the ball bounces back to a height of $h_1 = 10\left(\frac{3}{4}\right)$, comes down and hits the floor for the second time. The total distance it has traveled is

$$D_2 = h_0 + 2h_1 = 10 + 2 \cdot 10\left(\frac{3}{4}\right) = 25 \text{ feet}.$$

Then the ball bounces back to a height of $h_2 = 10\left(\frac{3}{4}\right)^2$, comes down and hits the floor for the third time. It has traveled

$$D_3 = h_0 + 2h_1 + 2h_2 = 10 + 2 \cdot 10\left(\frac{3}{4}\right) + 2 \cdot 10\left(\frac{3}{4}\right)^2 = 25 + 2 \cdot 10\left(\frac{3}{4}\right)^2 = 36.25 \text{ feet}.$$

Similarly,

$$D_4 = h_0 + 2h_1 + 2h_2 + 2h_3$$
$$= 10 + 2 \cdot 10\left(\frac{3}{4}\right) + 2 \cdot 10\left(\frac{3}{4}\right)^2 + 2 \cdot 10\left(\frac{3}{4}\right)^3$$
$$= 36.25 + 2 \cdot 10\left(\frac{3}{4}\right)^3$$
$$\approx 44.688 \text{ feet}.$$

(c) When the ball hits the floor for the $n^{\text{th}}$ time, its last bounce was of height $h_{n-1}$. Thus, by the method used in part (b), we get

$$D_n = h_0 + 2h_1 + 2h_2 + 2h_3 + \cdots + 2h_{n-1}$$

$$= 10 + \underbrace{2 \cdot 10 \left(\frac{3}{4}\right) + 2 \cdot 10 \left(\frac{3}{4}\right)^2 + 2 \cdot 10 \left(\frac{3}{4}\right)^3 + \cdots + 2 \cdot 10 \left(\frac{3}{4}\right)^{n-1}}_{\text{finite geometric series}}$$

$$= 10 + 2 \cdot 10 \cdot \left(\frac{3}{4}\right) \left(1 + \left(\frac{3}{4}\right) + \left(\frac{3}{4}\right)^2 + \cdots + \left(\frac{3}{4}\right)^{n-2}\right)$$

$$= 10 + 15 \left(\frac{1 - \left(\frac{3}{4}\right)^{n-1}}{1 - \left(\frac{3}{4}\right)}\right)$$

$$= 10 + 60 \left(1 - \left(\frac{3}{4}\right)^{n-1}\right).$$

## CHECK YOUR UNDERSTANDING

1. True. $a_1 = (1)^2 + 1 = 2$.

5. True. The differences between successive terms are all 1.

9. True. The first partial sum is just the first term of the sequence.

13. True. The sum is $n$ terms of 3. That is, $3 + 3 + \cdots + 3 = 3n$.

17. False. The terms of the series can be negative so partial sums can decrease.

21. True. If $a = 1$ and $r = -\frac{1}{2}$, it can be written $\sum_{i=0}^{5} \left(-\frac{1}{2}\right)^i$.

25. False. If payments are made at the end of each year, after 20 years, the balance at 5% is about $66,000, while at 10% it would be about $115,000. If payments are made at the start of each year, the corresponding figures are $69,000 and $126,000.

29. False. The series does not converge since the odd terms ($Q_1$, $Q_3$, etc.) are all $-1$.

33. False. An arithmetic series with $d \neq 0$ diverges.

# CHAPTER FOURTEEN

## Solutions for Section 14.1

### Exercises

**1.** We use a parameter $t$ so that when $t = 0$ we have $x = 1$ and $y = 3$. One possible parameterization is $x = 1 + 2t$, $y = 3 + t, 0 \leq t \leq 1$.

**5.** True. Eliminating $t$ gives $x = 3(y - 3)$. Thus we have the straight line $y = 3 + \frac{x}{3}$. We must also check the end points $t = 0$ and $t = 1$. When $t = 0$ we have $x = 0$ and $y = 3$, and when $t = 1$ we have $x = 3$ and $y = 3 + 1 = 4$.

**9.** The graph of the parametric equations is in Figure 14.1.

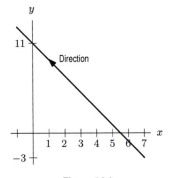

**Figure 14.1**

It is given that $x = 5 - 2t$, thus $t = (5 - x)/2$. Substitute this into the second equation:

$$y = 1 + 4t$$
$$y = 1 + 4\frac{(5 - x)}{2}$$
$$y = 11 - 2x.$$

**13.** The graph of the parametric equations is in Figure 14.2.

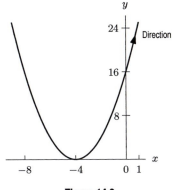

**Figure 14.2**

Since $x = t - 3$, we have $t = x + 3$. Substitute this into the second equation:

$$y = t^2 + 2t + 1$$
$$y = (x + 3)^2 + 2(x + 3) + 1$$
$$= x^2 + 8x + 16 = (x + 4)^2.$$

**17.** The graph of the parametric equations is in Figure 14.3.

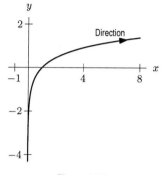

**Figure 14.3**

Since $x = t^3$, we take the natural log of both sides and get $\ln x = 3 \ln t$ or $\ln t = 1/3 \ln x$. We are given that $y = 2 \ln t$, thus,

$$y = 2 \left( \frac{1}{3} \ln x \right) = \frac{2}{3} \ln x.$$

**21.** This is like Example 5 on page 573 of the text, except that the $x$-coordinate goes all the way to 2 and back. So the particle traces out the rectangle shown in Figure 14.4.

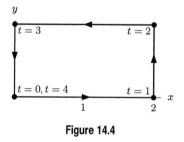

**Figure 14.4**

## Problems

**25.** The particle moves clockwise: For $0 \le t \le \frac{\pi}{2}$, starting at $(1, 0)$ as $t$ increases, we have $x = \cos t$ decreasing and $y = -\sin t$ decreasing. Similarly, for the time intervals $\frac{\pi}{2} \le t \le \pi, \pi \le t \le \frac{3\pi}{2}$, and $\frac{3\pi}{2} \le t \le 2\pi$, we see that the particle moves clockwise. The same is true for all $-\infty < t < +\infty$.

**29.** In all three cases, $y = x^2$, so that the motion takes place on the parabola $y = x^2$.

In case (a), the $x$-coordinate always increases at a constant rate of one unit distance per unit time, so the equations describe a particle moving to the right on the parabola at constant horizontal speed.

In case (b), the $x$-coordinate is never negative, so the particle is confined to the right half of the parabola. As $t$ moves from $-\infty$ to $+\infty$, $x = t^2$ goes from $\infty$ to $0$ to $\infty$. Thus the particle first comes down the right half of the parabola, reaching the origin $(0,0)$ at time $t = 0$, where it reverses direction and goes back up the right half of the parabola.

In case (c), as in case (a), the particle traces out the entire parabola $y = x^2$ from left to right. The difference is that the horizontal speed is not constant. This is because a unit change in $t$ causes larger and larger changes in $x = t^3$ as $t$ approaches $-\infty$ or $\infty$. The horizontal motion of the particle is faster when it is farther from the origin.

**33.** For $0 \leq t \leq 2\pi$, the graph is in Figure 14.5.

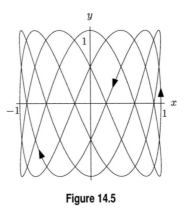

**Figure 14.5**

**37.** **(a)** Since the $x$-coordinate and the $y$-coordinate are always the same (they both equal $t$) , the bug follows the path $y = x$.
   **(b)** The bug starts at $(1, 0)$ because $\cos 0 = 1$ and $\sin 0 = 0$. Since the $x$-coordinate is $\cos x$, and the $y$-coordinate is $\sin x$, the bug follows the path of a unit circle, traveling counterclockwise. It reaches the starting point of $(1, 0)$ when $t = 2\pi$, because $\sin t$ and $\cos t$ are periodic with period $2\pi$.
   **(c)** Now the $x$-coordinate varies from $1$ to $-1$, while the $y$-coordinate varies from $2$ to $-2$; otherwise, this is much like part (b) above. If we plot several points, the path looks like an ellipse, which is a circle stretched out in one direction.

**41.** The particle moves back and forth between $-1$ and $1$. See Figure 14.6.

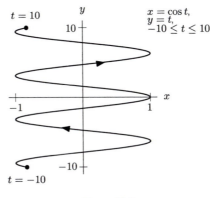

**Figure 14.6**

## Solutions for Section 14.2

### Exercises

**1.** Explicit. For each $x$ we can write down the corresponding values of $y$.

**5.** Implicit. This is a quadratic in $y$ so solving for $y$ by completing the square, we obtain $y = 1+\sqrt{2+x}$ and $y = 1-\sqrt{2+x}$.

**9.** Multiplying by $(y-4)^2$ gives

$$4 - (x-4)^2 - (y-4)^2 = 0$$
$$(x-4)^2 + (y-4)^2 = 4.$$

Thus, the center is $(4,4)$, and the radius is $2$.

**13.** We can use $x = 5\sin t$, $y = -5\cos t$ for $0 \le t \le 2\pi$.

**17.** We consider $x = -2 + \sqrt{5}\sin t$, $y = 1 + \sqrt{5}\cos t$ for $0 \le t \le 2\pi$. However, as $t$ increases from $0$, the value of $x$ increases, so this starts at the correct point but goes clockwise. We can use

$$x = -2 - \sqrt{5}\sin t, \quad y = 1 + \sqrt{5}\cos t, \quad \text{for } 0 \le t \le 2\pi.$$

### Problems

**21. (a)** Center is $(2,-4)$ and radius is $\sqrt{20}$.

   **(b)** Rewriting the original equation and completing the square, we have

$$2x^2 + 2y^2 + 4x - 8y = 12$$
$$x^2 + y^2 + 2x - 4y = 6$$
$$(x^2 + 2x + 1) + (y^2 - 4y + 4) - 5 = 6$$
$$(x+1)^2 + (y-2)^2 = 11.$$

So the center is $(-1,2)$, and the radius is $\sqrt{11}$.

**25.** Since $y - 3 = \sin t$ and $x = 4\sin^2 t$, this parameterization traces out the parabola $x = 4(y-3)^2$ for $2 \le y \le 4$.

**29.** Explicit: $y = \sqrt{4 - x^2}$
   Implicit: $y^2 = 4 - x^2$ or $x^2 + y^2 = 4$, $y > 0$
   Parametric: $x = 4\cos t$, $y = 4\sin t$, with $0 \le t \le \pi$.

## Solutions for Section 14.3

### Exercises

**1. (a)** The center is at the origin. The diameter in the $x$-direction is $4$ and the diameter in the $y$-direction is $2\sqrt{5}$.

   **(b)** The equation of the ellipse is

$$\frac{x^2}{2^2} + \frac{y^2}{(\sqrt{5})^2} = 1 \quad \text{or} \quad \frac{x^2}{4} + \frac{y^2}{5} = 1.$$

**5.** We consider $x = -2\cos t$, $y = 5\sin t$, for $0 \le t \le 2\pi$. However, in this parameterization, $y$ increases as $t$ increases from $0$, so it traces clockwise. Thus, we take $x = -2\cos t$, $y = -5\sin t$, for $0 \le t \le 2\pi$.

**9.** The fact that the parameter is called $s$, not $t$, makes no difference. The minus sign means that the ellipse is traced out in the opposite direction. The graph of the ellipse in the $xy$-plane is the same as the ellipse in the example, and it is traced out once as $s$ increases from $0$ to $2\pi$.

## Problems

**13.** Completing the square on $x^2 - 2x$ and $y^2 + 4y$:

$$\frac{1}{4}(x^2 - 2x) + y^2 + 4y + \frac{13}{4} = 0$$

$$\frac{1}{4}((x-1)^2 - 1) + (y+2)^2 - 4 + \frac{13}{4} = 0$$

$$\frac{1}{4}(x-1)^2 - \frac{1}{4} + (y+2)^2 - 4 + \frac{13}{4} = 0$$

$$\frac{(x-1)^2}{4} + (y+2)^2 = 1.$$

The center is $(1, -2)$, and $a = 2$, $b = 1$.

**17.** Factoring out the 9 from $9x^2 + 9x = 9(x^2 + x)$ and the 4 from $4y^2 - 4y = 4(y^2 - y)$ and completing the square on $x^2 + x$ and $y^2 - y$:

$$9(x^2 + x) + 4(y^2 - y) = \frac{131}{4}$$

$$9\left(\left(x + \frac{1}{2}\right)^2 - \frac{1}{4}\right) + 4\left(\left(y - \frac{1}{2}\right)^2 - \frac{1}{4}\right) = \frac{131}{4}$$

$$9\left(x + \frac{1}{2}\right)^2 - \frac{9}{4} + 4\left(y - \frac{1}{2}\right)^2 - 1 = \frac{131}{4}$$

$$9\left(x + \frac{1}{2}\right)^2 + 4\left(y - \frac{1}{2}\right)^2 = \frac{144}{4} = 36.$$

Dividing by 36 to get 1 on the right:

$$\frac{9\left(x + \frac{1}{2}\right)^2}{36} + \frac{4\left(y - \frac{1}{2}\right)^2}{36} = \frac{36}{36}$$

$$\frac{\left(x + \frac{1}{2}\right)^2}{4} + \frac{\left(y - \frac{1}{2}\right)^2}{9} = 1.$$

The center is $\left(-\frac{1}{2}, \frac{1}{2}\right)$, and $a = 2$, $b = 3$.

**21. (a)** Clearing the denominator, we have

$$r(1 - \epsilon \cos \theta) = r_0$$

$$r - r\epsilon \cos \theta = r_0$$

$$r - \epsilon x = r_0 \qquad \text{because } x = r \cos \theta$$

$$r = \epsilon x + r_0$$

$$r^2 = (\epsilon x + r_0)^2 \qquad \text{squaring both sides}$$

$$x^2 + y^2 = \epsilon^2 x^2 + 2\epsilon r_0 x + r_0^2 \qquad \text{because } r^2 = x^2 + y^2$$

$$x^2 - \epsilon^2 x^2 - 2\epsilon r_0 x + y^2 = r_0^2 \qquad \text{regrouping}$$

$$(1 - \epsilon^2)x^2 - 2\epsilon r_0 x + y^2 = r_0^2 \qquad \text{factoring}$$

$$Ax^2 - Bx + y^2 = r_0^2,$$

where $A = 1 - \epsilon^2$ and $B = 2\epsilon r_0$. From Question 20, we see that this is the formula of an ellipse.

**(b)** In the fraction $r_0/(1 - \epsilon \cos \theta)$, the numerator is a constant. Thus, the fraction's value is largest when the denominator is smallest, and smallest when the denominator is largest. The largest value of the denominator occurs at $\cos \theta = -1$, that is, at $\theta = \pi$, and so the minimum value of $r$ is

$$r = \frac{r_0}{1 - \epsilon \cos \pi} = \frac{r_0}{1 + \epsilon}.$$

The smallest value of the denominator occurs at $\cos\theta = 1$, that is, at $\theta = 0$, and so the minimum value of $r$ is

$$r = \frac{r_0}{1 - \epsilon\cos 0} = \frac{r_0}{1 - \epsilon}.$$

(c) See Figure 14.7. We know from part (b) that at $\theta = 0$, $r = r_0/(1 - \epsilon) = 6/(1 - 1/2) = 12$, and that at $\theta = \pi$, $r = r_0/(1+\epsilon) = 6/(1+1/2) = 4$. In addition, at $\theta = \pi/2$ and $\theta = 3\pi/2$, we see that $\cos\theta = 0$ and so $r = r_0 = 6$. This gives us the four points labeled in the figure. By symmetry, the center of the ellipse must lie between the points $(4, \pi)$ and $(12, 0)$, or, in Cartesian coordinates, the points $(-4, 0)$ and $(12, 0)$. Thus, the center is at $(8, 0)$.

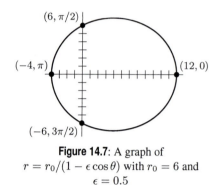

**Figure 14.7**: A graph of
$r = r_0/(1 - \epsilon\cos\theta)$ with $r_0 = 6$ and
$\epsilon = 0.5$

(d) The length of the horizontal axis is given by the minimum value of $r$ (at $\theta = \pi$) plus the maximum value of $r$ (at $\theta = 0$):

$$\text{Horizontal axis length} = r_{\min} + r_{\max}$$
$$= \frac{r_0}{1 + \epsilon} + \frac{r_0}{1 - \epsilon}$$
$$= \frac{r_0(1 - \epsilon) + r_0(1 + \epsilon)}{(1 - \epsilon)(1 + \epsilon)}$$
$$= \frac{2r_0}{1 - \epsilon^2}.$$

(e) At $\epsilon = 0$, we have $r = r_0/(1 - 0) = r_0$, which is the equation (in polar coordinates) of a circle of radius $r_0$. In our formula in part (d), we see that as $\epsilon$ gets closer to 1, the denominator gets closer to 0, and so the length of the horizontal axis increases rapidly. However, the $y$-intercepts at $(r_0, \pi/2)$ and $(r_0, 3\pi/2)$ do not change. For this to be true, the ellipse must be getting longer and longer along the horizontal axis as the eccentricity gets close to 1, but not be changing very much along the vertical axis.

## Solutions for Section 14.4

### Exercises

1. (a) The vertices are at $(0, 7)$ and $(0, -7)$. The center is at the origin.
   (b) The asymptotes have slopes $7/2$ and $-7/2$. The equations of the asymptotes are

$$y = \frac{7}{2}x \quad \text{and} \quad y = -\frac{7}{2}x.$$

   (c) The equation of the hyperbola is

$$\frac{y^2}{7^2} - \frac{x^2}{2^2} = 1 \quad \text{or} \quad \frac{y^2}{49} - \frac{x^2}{4} = 1.$$

**5.** The hyperbola is centered at the origin, and $a = 2$, $b = 7$. We can use $x = 2\tan t$, $y = 7\sec t = 7/\cos t$.

If $0 < t < \pi/2$, then $x > 0$, $y > 0$, so we have Quadrant I.

If $\pi/2 < t < \pi$, then $x < 0$, $y < 0$, so we have Quadrant III.

If $\pi < t < 3\pi/2$, then $x > 0$, $y < 0$, so we have Quadrant IV.

If $3\pi/2 < t < 2\pi$, then $x < 0$, $y > 0$, so we have Quadrant II.

So the upper half is given by $0 \le t < \pi/2$ together with $3\pi/2 < t < 2\pi$.

## Problems

**9.** Factoring out $-1$ from $-y^2 + 4y = -(y^2 - 4y)$ and completing the square on $x^2 - 2x$ and $y^2 - 4y$ gives

$$\frac{1}{4}(x^2 - 2x) - (y^2 - 4y) = \frac{19}{4}$$

$$\frac{1}{4}((x-1)^2 - 1) - ((y-2)^2 - 4) = \frac{19}{4}$$

$$\frac{(x-1)^2}{4} - \frac{1}{4} - (y-2)^2 + 4 = \frac{19}{4}$$

$$\frac{(x-1)^2}{4} - (y-2)^2 = 1.$$

The center is $(1, 2)$, the hyperbola opens right-left, and $a = 2$, $b = 1$.

**13.** Factoring out 4 from $4x^2 - 8x = 4(x^2 - 2x)$ and 36 from $36y^2 - 36y = 36(y^2 - y)$ and completing the square on $x^2 - 2x$ and $y^2 - y$ gives

$$4(x^2 - 2x) = 36(y^2 - y) - 31$$

$$4((x-1)^2 - 1) = 36\left(\left(y - \frac{1}{2}\right)^2 - \frac{1}{4}\right) - 31$$

$$4(x-1)^2 - 4 = 36\left(y - \frac{1}{2}\right)^2 - 9 - 31$$

$$4(x-1)^2 = 36\left(y - \frac{1}{2}\right)^2 - 36.$$

Moving $36(y - \frac{1}{2})^2$ to the left and dividing by $-36$ to get 1 on the right:

$$\frac{4(x-1)^2}{-36} - \frac{36\left(y - \frac{1}{2}\right)^2}{-36} = -\frac{36}{-36}$$

$$-\frac{(x-1)^2}{9} + \left(y - \frac{1}{2}\right)^2 = 1$$

$$\left(y - \frac{1}{2}\right)^2 - \frac{(x-1)^2}{9} = 1.$$

The center is $(1, \frac{1}{2})$, the hyperbola opens up-down, and $a = 3$, $b = 1$.

**17. (a)** The center is $(-5, 2)$, vertices are $(-5 + \sqrt{6}, 2)$ and $(-5 - \sqrt{6}, 2)$. The asymptotes are $y = \pm\frac{2}{\sqrt{6}}(x + 5) + 2$. Figure 14.8 shows the hyperbola.

**(b)** Rewriting the equation and competing the square, we have

$$x^2 - y^2 + 2x = 4y + 17$$

$$x^2 - y^2 + 2x - 4y = 17$$

$$(x^2 + 2x + 1) - (y^2 + 4y + 4) + 3 = 17$$

$$(x+1)^2 - (y+2)^2 = 14$$

$$\frac{(x+1)^2}{14} - \frac{(y+2)^2}{14} = 1.$$

Thus the center is $(-1, -2)$; vertices are $(-1 - \sqrt{14}, -2)$ and $(-1 + \sqrt{14}, -2)$; asymptotes are $y = \pm(x+1) - 2$, that is, $y = x - 1$ and $y = -x - 3$. Figure 14.9 shows the hyperbola.

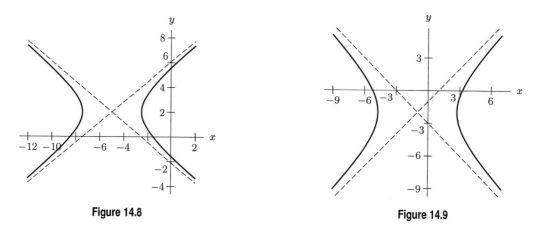

Figure 14.8                                                        Figure 14.9

## Solutions for Section 14.5

### Exercises

1. This is the standard form of an ellipse. Since $a > b$ the major axis is the $x$-axis.

5. Subtract the $x^2$ term from both sides to get the standard form of a hyperbola. Since the $y^2$ term is positive, the vertices are on the $y$-axis.

9. The two focal points lie on the major axis, and are equidistant from the center. Therefore, the major axis is vertical and the other focal point is at $(0, -2)$.

13. In the form
$$\frac{x^2}{a^2} - \frac{y^2}{b^2} = 1$$
we see that $a^2 = b^2 = 1$. The focal points lie on the same axis as the positive squared term. Therefore the focal points are at $(\pm c, 0)$, where $c = \sqrt{a^2 + b^2} = \sqrt{1 + 1} = \sqrt{2}$. The two focal points are $(\sqrt{2}, 0)$ and $(-\sqrt{2}, 0)$.

### Problems

17. The reflective properties of a parabola are somewhat helpful, because the transmission can be focused to go in only one direction. However, the location of the transmission will be perfectly clear to anyone noticing the transmission. A series of elliptical mirrors would be much more complicated and expensive to build, and by following the signal, someone could figure out where the transmission was coming from without your being aware, because surveying a series of mirrors is complicated. A hyperbolic reflection allows you to survey a small mirror near your camp while at the same time making it appear that your camp is somewhere else (at the other focus of the hyperbola). So you should follow the mess sergeant's advice.

21. From the figure, we see the vertex has coordinates $(1, 0)$ and the equation must be of the form
$$x = ay^2 + 1$$
The distance, $c$ from the vertex to the focus is 2. Using $c = 1/(4a)$, we solve for $a$ in $2 = 1/(4a)$ and find $a = 1/8$. The equation is $x = (1/8)y^2 + 1$. The directrix is the line $x = -1$, since it is 2 units to the left of the vertex.

**25.** Rewrite the formula as
$$(x^2 + 4x) + 2(y^2 - 6y) = 3.$$
Complete the square of both the $x$ and $y$ expressions
$$(x^2 + 4x + 4) + 2(y^2 - 6y + 9) = 3 + 4 + 2(9),$$
and simplify to
$$(x + 2)^2 + 2(y - 3)^2 = 25$$
or
$$\frac{(x + 2)^2}{25} + \frac{(y - 3)^2}{(25/2)} = 1.$$

This is an ellipse with center at $(-2, 3)$, and $a = 5$ and $b = \sqrt{25/2}$. Since $a > b$, the distance $c$, from the center to a focus point is found from
$$c = \sqrt{a^2 - b^2} = \sqrt{25 - 25/2} = \sqrt{25/2} = 5/\sqrt{2}.$$
With $c = 5/\sqrt{2}$ the focal points are at $(-2 + 5/\sqrt{2}, 3)$ and $(-2 - 5/\sqrt{2}, 3)$.

**29.** First rewrite the equation in the form,
$$\frac{x^2}{8} - \frac{y^2}{16} = 1.$$

Because the $x^2$ term is positive and the $y^2$ term is negative this hyperbola has vertices on a horizontal line through its center. The center is at the origin. The length from the center to the vertex is the distance $a = 2\sqrt{2}$ which is read from the equation since $a^2 = 8$. Thus the vertices are at the points $(2\sqrt{2}, 0)$ and $(-2\sqrt{2}, 0)$. Since $b^2 = 16$ and $a^2 = 8$, we find $c^2 = a^2 + b^2 = 24$, and $c = \sqrt{24}$. The focal points are $(\sqrt{24}, 0)$ and $(-\sqrt{24}, 0)$.

**33.** After passing through the second focal point it is reflected back to the original focal point.

**37.** For a cross-section of the dish centered at the origin, the equation of the parabola is $y = ax^2$. We know that the rim of the dish is the point $(12, 4)$, thus $4 = a(12)^2$ and solving for $a$ we find $a = 1/36$. Using the equation $c = 1/(4a)$, we find
$$c = \frac{1}{4(1/36)} = 9.$$

The end of the arm is placed 9 inches above the center of the dish to be the focus of the incoming signal.

**41.** **(a)** The major axis has length $88 + 5250 = 5338$ million km.
**(b)** Since $2a = 5338$, we find $a = 2669$. Let $(\pm c, 0)$ be the two focal points. Since $a = 2669 = c + 88$, we find $c = 2581$. To find $b$ we use the formula $c^2 = a^2 - b^2$ and find $b^2 = 2669^2 - 2581^2 = 462,000$, so $b = 680$.
   Shifting the axis so the sun is at the origin we have,
$$\frac{(x - 2581)^2}{2669^2} + \frac{y^2}{680^2} = 1.$$
**(c)** Use the parametric equations of an ellipse centered at $(h, k)$,
$$x = h + a\cos t, \quad y = k + b\sin t, \quad 0 \le t \le 2\pi$$
to obtain
$$x = 2581 + 2669\cos t, \quad y = 680\sin t, \quad 0 \le t \le 2\pi.$$

**45.** Consider the set of points $(x, y)$ so that the difference of the distances to two focal points $(\pm c, 0)$ is constant. Note that the $x$-intercepts $(\pm a, 0)$ satisfy this condition.
   Using the distance formula, we see that the distance from $(x, y)$ to $(c, 0)$ is $\sqrt{(x - c)^2 + y^2}$ and the distance from $(x, y)$ to $(-c, 0)$ is $\sqrt{(x + c)^2 + y^2}$. The distance from $(a, 0)$ to $(-c, 0)$ is $c + a$ and the distance from $(a, 0)$ to $(c, 0)$ is $c - a$ since $c > a$. We have:

Difference of distances from $(x, y)$ to focal points = Difference of distances from $(a, 0)$ to focal points
$$\sqrt{(x - c)^2 + y^2} - \sqrt{(x + c)^2 + y^2} = (c + a) - (c - a)$$
$$\sqrt{(x - c)^2 + y^2} - \sqrt{(x + c)^2 + y^2} = 2a$$
$$\sqrt{(x - c)^2 + y^2} = 2a + \sqrt{(x + c)^2 + y^2}.$$

We square both sides and simplify:

$$(x - c)^2 + y^2 = 4a^2 + 4a\sqrt{(x+c)^2 + y^2} + ((x+c)^2 + y^2)$$
$$x^2 - 2cx + c^2 + y^2 = 4a^2 + 4a\sqrt{(x+c)^2 + y^2} + x^2 + 2cx + c^2 + y^2.$$

We cancel $x^2$, $c^2$, and $y^2$ from both sides, and subtract $2cx$ from both sides to obtain:

$$-4cx = 4a^2 + 4a\sqrt{(x+c)^2 + y^2}.$$

We divide through by 4:

$$-cx = a^2 + a\sqrt{(x+c)^2 + y^2},$$

and then isolate the square root:

$$-a\sqrt{(x+c)^2 + y^2} = a^2 + cx.$$

We square both sides again to obtain:

$$a^2((x+c)^2 + y^2) = a^4 + 2cxa^2 + c^2x^2$$
$$a^2(x^2 + 2cx + c^2 + y^2) = a^4 + 2cxa^2 + c^2x^2$$
$$a^2x^2 + 2cxa^2 + a^2c^2 + a^2y^2 = a^4 + 2cxa^2 + c^2x^2$$
$$a^2x^2 + a^2c^2 + a^2y^2 = a^4 + c^2x^2.$$

Since $c > a$, there is a positive number $b$ such that $c^2 = a^2 + b^2$:

$$a^2x^2 + a^2(a^2 + b^2) + a^2y^2 = a^4 + (a^2 + b^2)x^2$$
$$a^2x^2 + a^4 + a^2b^2 + a^2y^2 = a^4 + a^2x^2 + b^2x^2$$
$$a^2b^2 + a^2y^2 = b^2x^2$$
$$b^2x^2 - a^2y^2 = a^2b^2.$$

Dividing through by $a^2b^2$, we arrive at the equation for a hyperbola given in Section 14.3:

$$\frac{b^2x^2}{a^2b^2} - \frac{a^2y^2}{a^2b^2} = \frac{a^2b^2}{a^2b^2}$$
$$\frac{x^2}{a^2} - \frac{y^2}{b^2} = 1.$$

This is the equation for a hyperbola.

## Solutions for Section 14.6

### Exercises

**1.** We use $x = \sinh t$, $y = \cosh t$, for $-\infty < t < \infty$.

**5.** We use $x = 1 + 2\sinh t$, $y = -1 - 3\cosh t$, for $-\infty < t < \infty$.

**9.** Divide by 36 to rewrite the equation as

$$\frac{(y+3)^2}{4} - \frac{(x+1)^2}{9} = 1, \quad y > -3,$$

and use $x = -1 + 3\sinh t$, $y = -3 + 2\cosh t$, for $-\infty < t < \infty$.

**13.** The graph of $\sinh x$ in the text suggests that

$$\text{As } x \to \infty, \qquad \sinh x \to \tfrac{1}{2}e^x$$
$$\text{As } x \to -\infty, \qquad \sinh x \to -\tfrac{1}{2}e^{-x}.$$

Using the facts that

$$\text{As } x \to \infty, \qquad e^{-x} \to 0,$$
$$\text{As } x \to -\infty, \qquad e^{x} \to 0,$$

we can predict the same results algebraically:

$$\text{As } x \to \infty, \qquad \sinh x = \tfrac{e^x - e^{-x}}{2} \to \tfrac{1}{2}e^x$$
$$\text{As } x \to -\infty, \qquad \sinh x = \tfrac{e^x - e^{-x}}{2} \to -\tfrac{1}{2}e^{-x}.$$

## Problems

**17.** Factoring out 2 from $2x^2 - 12x = 2(x^2 - 6x)$ and 4 from $4y^2 + 4y = 4(y^2 + y)$ and completing the square on $x^2 - 6x$ and $y^2 + y$ gives

$$25 + 2(x^2 - 6x) = 4(y^2 + y), \quad y > -\frac{1}{2}$$

$$25 + 2((x-3)^2 - 9) = 4\left(\left(y + \frac{1}{2}\right)^2 - \frac{1}{4}\right), \quad y > -\frac{1}{2}$$

$$25 + 2(x-3)^2 - 18 = 4\left(y + \frac{1}{2}\right)^2 - 1, \quad y > -\frac{1}{2}$$

$$2(x-3)^2 = 4\left(y + \frac{1}{2}\right)^2 - 8, \quad y > -\frac{1}{2}.$$

Then, moving $4(y + \frac{1}{2})^2$ to the left and dividing by $-8$ to get 1 on the right,

$$\frac{2(x-3)^2}{-8} - \frac{4\left(y + \frac{1}{2}\right)^2}{-8} = 1, \quad y > -\frac{1}{2}$$

$$\frac{\left(y + \frac{1}{2}\right)^2}{2} - \frac{(x-3)^2}{4} = 1, \quad y > -\frac{1}{2},$$

so we use $x = 3 + 2\sinh t, y = -\frac{1}{2} + \sqrt{2}\cosh t$ for $-\infty < t < \infty$.

**21.** Yes. First, we observe that
$$\cosh 2x = \frac{e^{2x} + e^{-2x}}{2}.$$

Now, using the fact that $e^x \cdot e^{-x} = 1$, we calculate

$$\cosh^2 x = \left(\frac{e^x + e^{-x}}{2}\right)^2$$
$$= \frac{(e^x)^2 + 2e^x \cdot e^{-x} + (e^{-x})^2}{4}$$
$$= \frac{e^{2x} + 2 + e^{-2x}}{4}.$$

Similarly, we have

$$\sinh^2 x = \left(\frac{e^x - e^{-x}}{2}\right)^2$$

$$= \frac{(e^x)^2 - 2e^x \cdot e^{-x} + (e^{-x})^2}{4}$$
$$= \frac{e^{2x} - 2 + e^{-2x}}{4}.$$

Thus, to obtain $\cosh 2x$, we need to add (rather than subtract) $\cosh^2 x$ and $\sinh^2 x$, giving

$$\cosh^2 x + \sinh^2 x = \frac{e^{2x} + 2 + e^{-2x} + e^{2x} - 2 + e^{-2x}}{4}$$
$$= \frac{2e^{2x} + 2e^{-2x}}{4}$$
$$= \frac{e^{2x} + e^{-2x}}{2}$$
$$= \cosh 2x.$$

Thus, we see that the identity relating $\cosh 2x$ to $\cosh x$ and $\sinh x$ is

$$\cosh 2x = \cosh^2 x + \sinh^2 x.$$

**25.** We know that $\sinh(iz) = i \sin z$, where $z$ is real. Substituting $z = ix$, where $x$ is real so $z$ is imaginary, we have

$$\sinh(iz) = i \sin z$$
$$\sinh(i \cdot ix) = i \sin(ix) \qquad \text{substituting } z = ix$$
$$\sinh(-x) = i \sin(ix).$$

But $\sinh(-x) = -\sinh(x)$, thus we have

$$-\sinh x = i \sin(ix).$$

Multiplying both sides by $i$ gives

$$-i \sinh x = -1 \sin(ix).$$

Thus,

$$i \sinh x = \sin(ix).$$

# Solutions for Chapter 14 Review

## Exercises

**1.** The coefficients of $x^2$ and $y^2$ are equal, so this is a circle with center $(0, 3)$ and radius $\sqrt{5}$.

**5.** Dividing by 36 to get 1 on the right gives

$$\frac{9(x-5)^2}{36} + \frac{4y^2}{36} = \frac{36}{36}$$
$$\frac{(x-5)^2}{4} + \frac{y^2}{9} = 1.$$

This is an ellipse centered at $(5, 0)$, with $a = 2$, $b = 3$.

**9.** One possible answer is $x = 3\cos t, y = -3\sin t, 0 \le t \le 2\pi$.

**13.** The ellipse $x^2/25 + y^2/49 = 1$ can be parameterized by $x = 5\cos t, y = 7\sin t, 0 \le t \le 2\pi$.

**17.** In the form

$$\frac{x^2}{a^2} + \frac{y^2}{b^2} = 1$$

we see that $a > b$. Therefore the focal points are at $(\pm c, 0)$, where $c = \sqrt{a^2 - b^2} = \sqrt{25 - 4} = \sqrt{21}$. The two focal points are $(\sqrt{21}, 0)$ and $(-\sqrt{21}, 0)$.

## Problems

**21.** A parametric equation for the circle is

$$x = \cos t, y = \sin t.$$

As $t$ increases from 0, we have $x$ increasing and $y$ decreasing, which is a counterclockwise movement, so this parameterization is correct.

**25.** Factoring out 6 from $6x^2 - 12x = 6(x^2 - 2x)$ and 9 from $9y^2 + 6y = 9(y^2 + \frac{2}{3}y)$, and completing the square on $x^2 - 2x$ and $y^2 + \frac{2}{3}y$ gives

$$6(x^2 - 2x) - 9\left(y^2 + \frac{2}{3}y\right) + 1 = 0$$

$$6((x-1)^2 - 1) + 9\left(\left(y + \frac{1}{3}\right)^2 - \frac{1}{9}\right) + 1 = 0$$

$$6(x-1)^2 - 6 + 9\left(y + \frac{1}{3}\right)^2 - 1 + 1 = 0$$

$$6(x-1)^2 + 9\left(y + \frac{1}{3}\right)^2 = 6.$$

(Alternatively, you may recognize $9y^2 + 6y + 1$ as the perfect square $(3y+1)^2$). Dividing by 6 to get 1 on the right

$$(x-1)^2 + \frac{9\left(y + \frac{1}{3}\right)^2}{6} = 1$$

$$(x-1)^2 + \frac{3\left(y + \frac{1}{3}\right)^2}{2} = 1.$$

This is an ellipse centered at $(1, -\frac{1}{3})$ with $a = 1$, $b^2 = 2/3$, so $b = \sqrt{2/3}$.

**29. (a)** Since $P$ moves in a circle we have

$$x = 10\cos t$$
$$y = 10\sin t.$$

This completes a revolution in time $2\pi$.

**(b)** First, consider the planet as stationary at $(x_0, y_0)$. Then the equations for $M$ are

$$x = x_0 + 3\cos 8t$$

$$y = y_0 + 3\sin 8t.$$

The factor of 8 is inserted because for every $2\pi/8$ units of time, $8t$ covers $2\pi$, which is one orbit. But since $P$ moves, we must replace $(x_0, y_0)$ by the position of $P$. So we have

$$x = 10\cos t + 3\cos 8t$$

$$y = 10\sin t + 3\sin 8t.$$

(c) See Figure 14.10.

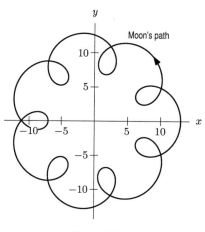

**Figure 14.10**

# CHECK YOUR UNDERSTANDING

**1.** True. The path is on the line $3x - 2y = 0$.

**5.** False. Starting at $t = 0$ the object is at $(0, 1)$. For $0 \leq t \leq \pi$, $x = \sin(t/2)$ increases and $y = \cos(t/2)$ decreases, thus the motion is clockwise.

**9.** False. The standard form of the circle is $(x + 4)^2 + (y + 5)^2 = 141$. The center is $(-4, -5)$.

**13.** False. There are many parameterizations; $x = \cos t, y = \sin t$ and $x = \sin t, y = \cos t$ are two of them.

**17.** True. Since

$$\frac{x^2}{4} + y^2 = \frac{(2 \cos t)^2}{4} + (\sin t)^2 = \cos^2 t + \sin^2 t = 1,$$

these equations parameterize the ellipse $(x^2/4) + y^2 = 1$.

**21.** False. It is defined as $\sinh x = \dfrac{e^x - e^{-x}}{2}$.

**25.** True. We have $\cosh(-x) = (e^{-x} + e^{-(-x)})/2 = (e^x + e^{-x})/2 = \cosh x$, so the hyperbolic cosine is an even function.

**29.** False. We have

$$\sinh \pi = \frac{e^\pi}{2} - \frac{e^{-\pi}}{2} = 11.549 \neq 0.$$

We do have $\sin \pi = 0$.

**33.** False. The two asymptotes are entirely in the region between the two branches of the hyperbola. The two focal points are outside this region. To move from an asymptote to a focal point you must cross the hyperbola.